M000311804

FOSSIL FUELS
IMPROVE THE PLANET

BY ALEX EPSTEIN

Permission requests for college or textbook use should be addressed to CIPress, 25592 La Mirada Street, Laguna Hills, CA 92653.

Printed in the United States of America

Book design by Marianne Epstein

First Printing, 2013

ISBN 978-0-9893448-0-7

TABLE OF CONTENTS

INTRODUCTION

I am the founder of the Center for Industrial Progress.

Center for Industrial Progress (CIP) is a for-profit think-tank seeking to bring about a new industrial revolution. We believe that human beings have the untapped potential to radically improve our lives by using technology to improve the planet across a multitude of industries: mining, manufacturing, agriculture, chemistry, and energy. Every individual has the potential for a longer, happier, healthier, safer, more comfortable, more meaningful, more opportunity-filled life.

The keys to a new industrial revolution are a new industrial philosophy, a new industrial policy, and a new approach to communication.

Philosophically, the so-called environmentalist movement has put forward the myth that a better environment means "saving" the planet from human industry. CIP shows that a better environment means improving the planet through human industry.

Politically, the so-called environmentalist movement has put forward the myth that a better environment means authoritarian control by environmental bureaucrats who prioritize sloths over human beings. CIP shows that a better environment means clear, scientific laws that protect both the right to develop one's property and the right to preserve clean air and water on one's property.

In communication, the so-called environmentalist movement has historically taken the moral high ground against industry by pretending that minimizing the "footprint" of industry is an ideal that will bring about a better, healthier world. CIP's aspirational approach to communication shows that the real ideal is industrial progress, the progressive improvement of the planet through technology and development—and has inspired thousands to change the way they think about industries such as energy, mining, and agriculture.

A crucial part of our mission is sharing our uniquely positive ideas and tactics with industry through training programs that inspire their employees and empower their communications teams. Only if industry starts to appreciate and articulate its true value, both economic and environmental, can we liberate industry from authoritarian government and bring about the next industrial revolution.

To learn more about how CIP can help your business or industry, email us at support@industrialprogress.net or visit www.industrialprogress.com/contact.

UNDERSTANDING FOSSIL FUELS

CHAPTER 1: FOSSIL FUELS IMPROVE THE PLANET

The basic question underlying our energy policy debates is this:

Should we be free to generate more and more energy using fossil fuels? Or should we restrict and progressively outlaw fossil fuels as "dirty energy"?

I believe that if we look at the big picture, the facts are clear. If we want a healthy, livable environment, then we must be free to use fossil fuels.

Why? Because for the foreseeable future, fossil fuels provide the key to a great environment: abundant, affordable, reliable energy.

We're taught in school that the key to a great environment is to minimize our "impact" on it. We think of our environment as something that starts out healthy and that we humans mess up. Not so. Nature does not give us a healthy environment to live in; until the fossil-fueled industrial revolution of the last two centuries, human beings lived in an environment that was low on useful resources and high on danger.[1]

Today's industrialized environment is the cleanest, healthiest in history. If you want to see what "dirty" looks like, go to a country that is still living in "natural," pre-industrial times. Try choking on the natural smoke of a natural open fire burning natural wood or animal dung—the kind of air pollution that has been almost eliminated by modern, centralized power plants. Try getting your water from a local brook that is naturally infested with the natural germs of all the local animals—the once-perennial threat that modern, fossil-fuel-powered water purification systems eliminate. Try coping with the dramatic temperature and weather swings that occur in nearly any climate—a threat that fossil-fuel powered air-conditioning, heating, and construction have made extremely rare.

We live in an environment where the air we breathe and the water we drink and the food we eat will not make us sick, and where we can cope with the often hostile climate of nature. That is a huge achievement—an achievement that lives or dies with the mass-production of energy. We can live this way only by getting high-powered machines to do the vast majority of our physical work for us.[2] Energy is what we need to build sturdy homes, to produce huge amounts of fresh food, to generate heat and air-conditioning, to irrigate deserts, to dry malaria-infested swamps, to build hospitals, to manufacture pharmaceuticals. And those of us who enjoy exploring the rest of nature should never forget that oil is what enables us to explore to our heart's content, which pre-industrial people didn't have the time, wealth, energy, or technology to do.

The more affordable, reliable energy we can produce, the better world we can build. In order for everyone in the world to have as much energy as the average German, we would need to produce twice as much energy.[3] 1.3 billion people in the world lack electricity: that means no light at night, no refrigeration, no factories, no water purification.[4] And all of us could do more with more energy to travel and with lower electric and heating bills.

So it is very upsetting to me to see the fossil fuel industry, by far the best producer of energy, attacked as a "dirty" industry to be eliminated. That is a policy of mass destruction. And unfortunately, it's not an innocent mistake; the "environmentalist" leaders who hate fossil fuels also hate nuclear power and hydroelectric power, the only other two sources that have provided any significant affordable, reliable power.

The vicious attack on practical energy is rationalized by a phony enthusiasm for solar and wind.[5] Don't fall for it. If solar or wind were good alternatives, they wouldn't need political advocates; they'd win out on the market. But solar and wind have been the biggest energy failure of the last century.

By trying to rely on unreliable, low-concentration streams of energy, they have produced unreliable, expensive power—which is why after decades of subsidies they produce less than half a percent of the world's energy, and all of it needs to be backed up by fossil fuels, nuclear, and hydro.[6] Some rich countries have tried to score moral points by paying exorbitant sums to buy and back up unreliable energy sources, and even they can't afford it. Industry runs on energy, and expensive energy means industrial decline. Germany's industrial electricity prices have doubled (and it would be more without the dozen new coal plants they are building) and "green" Spain has a youth unemployment rate of 50%.[7][8][9]

If "environmentalists" claimed that we must be forced to replace the steel girders in skyscrapers with wood girders, we would know that the result would be collapse. The same goes for replacing the best energy with the worst energy—the result would be the collapse of an entire civilization.

Are fossil fuels dirty? Fossil fuels have fueled the unprecedented industrial progress that doubled the human life expectancy and produced the cleanest, healthiest human environment in history.[10] That, unlike Al Gore's hysteria about 20-foot sea level rises, is a fact. It's also a fact that even though we hear so much hysteria about the one degree Celsius of climate change that has occurred since the Little Ice Age (half of it before major CO_2 emissions) so far, heat-related deaths keep going down and overall climate-related death rates have gone down 98% since we started using extremely large amounts of fossil fuels.[11] With technology powered by affordable, reliable energy, human beings can adapt to just about whatever happens in just about any climate. Without practical energy, no matter what the climate is, we're in trouble.

Fossil fuels are not "dirty energy." Fossil fuels are a health necessity to the human environment. What about the waste? We are incredibly spoiled and ungrateful if we call that "dirty." Every human and non-human activity creates waste products—certainly building monstrous solar and wind arrays out of hazardous materials does— but technology allows us to minimize dangerous waste. [12] [13]

The "dirty energy" objection is a dirty trick. Since everything creates some kind of waste byproduct, you can just oppose it by calling it "dirty." If you study the mining and the materials that go into solar panels and windmills, and the incredible amount of coal and oil that goes into manufacturing and transporting and assembling their parts, you can call them "dirty," too. [14] [15] [16] The "dirty" objection is just a convenient trick for people who really don't like any kind of industrial development—people who think that there's something unnatural and wrong about the modern, industrial way of life. Don't fall for it. Our nature as humans is to use our intelligence to build a better environment for ourselves. We should embrace fossil fuels, and embrace them with pride in the face of those who would destroy them. Remember this: No energy is dirtier than no energy.

CHAPTER 2: FOSSIL FUELS MAKE CATASTROPHES NON-CATASTROPHIC

One of the greatest and most unheralded successes of fossil-fuel-powered industrial capitalism is making our climate eminently livable.[17]

The mass-production of sturdy, weather-proof buildings ... the universal availability of heating and air conditioning ... the ability to flee the most vicious storms through modern transportation ... the protection from drought through modern irrigation and food transport ... the protection from disease through modern sanitation—all have been powered by fossil fuels. Combined, they have led to a 98% reduction in the number of climate-related deaths over the last century.[18] Given how obsessed America is about catastrophic global warming (or how some intellectuals/politicians want us to be), these facts should be well-known and incorporated into every discussion of industrial policy. Those who claim to care about a livable climate for the future should strive to understand the mechanisms by which industrial capitalism has already made our climate the most livable in history.

If they did so, they would learn from such thinkers as Ayn Rand and Ludwig Von Mises how capitalism, by permitting only voluntary associations among men, unleashes the individual human mind—and that billions of such minds, free to associate and trade however they choose, will engage in stupendously intricate, collaborative planning for everything from how to make sure they can always get groceries to how to account for nearly any weather contingency.

Armed with an understanding of individual freedom and individual planning, the climate-concerned would suspect that any preventable problem in dealing with weather—such as widespread vulnerability to flooding—is caused by government interference in voluntary trade, such as taxpayer-financed flood insurance that encourages people to live in high-flooding areas.

Unfortunately, an understanding of capitalism and climate is sorely lacking in the catastrophic global warming movement. A typical example is the ThinkProgress blog at the Center for American Progress (CAP). A typical post by one of its prestigious bloggers, Christian Parenti, captures the regrettable combination of arrogance and ignorance that most climate commentators exhibit.[19] The title of the post is designed to intimidate: "Climate Action Opponents Are Ensuring the Outcome They Claim to Oppose: Big Government."

A little translation is in order. From an individualistic perspective, "climate action" refers to the actions that free citizens take to make their climate as livable as possible—the kinds of actions that have eliminated the vast majority of climate danger since the mass-production of fossil fuels.

But from the perspective of CAP, which believes in extensive state control of the individual, "Climate Action" refers to dramatic restrictions on energy generated from hydrocarbons—the energy source that runs the industrial capitalist system that has increased our life expectancy from 30 to 80 years.[20]

How will banning the vast majority of modern energy production help us oppose "Big Government"? Because if we don't, the article argues, we would face so many catastrophic storms that the government would necessarily become a disaster-recovery Leviathan.

In addition to taking for granted that all warming is caused by human beings and that warming causes more storms (unlikely and baseless, respectively) Mr. Parenti takes as given that government is the only entity that can adapt to storms: "To adapt to climate change will mean coming together on a large scale and mobilizing society's full range of resources. In other words, Big Storms require Big Government."[21]

In fact, the larger-scale a problem, the more freedom, including the freedom to produce the best kinds of energy, is essential. As economist George Reisman brilliantly explains in his landmark essay on global warming economics:

> Even if global warming is a fact, the free citizens of an industrial civilization will have no great difficulty in coping with it—that is, of course, if their ability to use energy and to produce is not crippled by the environmental movement and by government controls otherwise inspired. The seeming difficulties of coping with global warming, or any other large-scale change, arise only when the problem is viewed from the perspective of government central planners.
>
> It would be too great a problem for government bureaucrats to handle (as is the production even of an adequate supply of wheat or nails, as the experience of the whole socialist world has so eloquently shown). But it would certainly not be too great a problem for tens and hundreds of millions of free, thinking individuals living under capitalism to solve. It would be solved by means of each individual being free to decide how best to cope with the particular aspects of global warming that affected him.
>
> Individuals would decide, on the basis of profit-and-loss calculations, what changes they needed to make in their businesses and in their personal lives, in order best to adjust to the situation. They would decide where it was now relatively more desirable to own land, locate farms and businesses, and live and work, and where it was relatively less desirable, and

what new comparative advantages each location had for the production of which goods. Factories, stores, and houses all need replacement sooner or later. In the face of a change in the relative desirability of different locations, the pattern of replacement would be different. Perhaps some replacements would have to be made sooner than otherwise. To be sure, some land values would fall and others would rise. Whatever happened, individuals would respond in a way that minimized their losses and maximized their possible gains. The essential thing they would require is the freedom to serve their self-interests by buying land and moving their businesses to the areas rendered relatively more attractive, and the freedom to seek employment and buy or rent housing in those areas.

Given this freedom, the totality of the problem would be overcome. This is because, under capitalism, the actions of the individuals, and the thinking and planning behind those actions, are coordinated and harmonized by the price system (as many former central planners of Eastern Europe and the former Soviet Union have come to learn). As a result, the problem would be solved in exactly the same way that tens and hundreds of millions of free individuals have solved greater problems than global warming, such as redesigning the economic system to deal with the replacement of the horse by the automobile, the settlement of the American West, and the release of the far greater part of the labor of the economic system from agriculture to industry. [22]

We should be thankful that previous generations were not governed by the regressive "progressive" philosophy that regards government coercion as the solution to future changes, whether economic or environmental. Had they followed it, we would have had the equivalent of Barack Obama or Christian Parenti dictating to millions of Americans when, how, or if they could transition to automobiles or go West or leave their farms. If we were really facing worse weather ahead, then nothing would be more important in preparing than more industry and more freedom.

CHAPTER 3: COAL IS HEALTHY

When the average American thinks of coal, what is the first word that comes to mind? The answer is almost certainly "dirty." As long as that's the case, our country is in trouble.

The common slogan "coal is dirty" is not just an empty slogan—it's a statement of a deep-seated idea in our culture, which is the idea that the use of coal harms our environment. That is a very powerful attack because environment is such an important issue.

It's connected to our health, the health of our loved ones, our quality of life, and our future. When people hear something is dirty, they want no part of it—and understandably so.

That's why, wherever you find coal losing, you'll find the "coal is dirty" idea.

- It's behind the EPA's effective ban on new coal power plants.

- It's behind regulators shutting down existing power plants.

- It's behind activists working to ban coal exports.

But if we look at the *overall* environmental impact of coal and other fossil fuels it's amazingly good.

To see coal's overall environmental impact, it's helpful to take a historical perspective.

Imagine we transported someone from 300 years ago, which was essentially a coal-free environment, to today's world, which has fundamentally been shaped by coal, oil, and natural gas. What would he think about our environment?

His reaction would be disbelief that such a clean, healthy environment was possible.

"How is this possible?" he would ask.

The air is so clean—where I come from we're breathing in smoke all day from the fire we need to burn in our furnace.

And the water. Everywhere I go there's this water that tastes so good, and it's all safe to drink, how is that possible? On my farm, we get our water from a brook we share with animals and my kids are always getting sick.

And then the weather. I mean, the weather isn't that much different, but you're so much safer in it; you guys can move a knob and make it cool when it's hot and warm when it's cold?

And then the food. I'm a farmer, but we only have a few different crops, and in the winter we can't grow anything so we eat the same bread over and over. But you guys are surrounded by all this amazing food—this is like the Garden of Eden.

And you have to tell me, what happened to all the disease? Where I'm from, we have insects all over the place giving us disease—my neighbor's son died of malaria—and you don't seem to have any of that here. What's your secret?

Well, if I were talking to him, I'd respond that the key was that we figured out how to produce cheap, plentiful, reliable energy from coal and other fossil fuels to build a modern, technological world that is healthier and safer than any world men have ever lived in.

Now as unbelievable as our environmental quality would be to him, I think there's one thing he would find more unbelievable: that in our culture we regard fossil fuel companies, especially coal companies, as villains.

"Why?"

"Well, there's a big anti-coal movement that says coal is dirty, and most people in our society agree with them."

"Wait, they say that the coal that helped make *this* environment is dirty? No, *where I come from is dirty*. This is not dirty. This is the cleanest thing I've ever seen. Whatever coal is, it doesn't make things dirty, it cleans things, it's healthy. I mean just look around."

And this visitor would be 100% right. Coal is healthy. It enables us to flourish.

Now, like anything, coal has certain safety challenges you want to minimize, but if you look at the big picture, it's incredibly positive.

We accept this principle readily in areas such as medicine. Take antibiotics. There are certain negatives to antibiotics, but overall they are incredibly beneficial—we wouldn't label them "dirty medicine." Why not take the same attitude toward coal?

The idea that "coal is dirty" is only plausible because we are taught to *define* coal by its negative environmental impacts—while ignoring its much greater environmental benefits. But to be objective, we need to define things by their overall impact—and if we do that, we see that coal is not dirty. It's healthy.

DEFENDING FOSSIL FUELS

CHAPTER 4: HOW THE COAL INDUSTRY SHOULD DEFEND ITSELF

Once you understand that coal and other fossil fuels improve our environment, your ability to defend them is incomparably greater.

Let's work through an example: the controversy over coal exports in the Pacific Northwest.

Here's a typical attack: "They're coming to ship their poison so they can poison the people in China. And that poison's going to come back here and poison your salmon and your children, so *don't let it happen*."[23] That was from Robert F. Kennedy, Jr.

So let's say you're debating Robert F. Kennedy Jr. in the media. How do you respond?

If you're clear that coal *improves* our environment, not just that it's less poisonous than he thinks, you can completely turn the tables and make clear that as supporters of coal *you're* the environmental benefactor and he's the environmental danger.

Here's how I might respond if I were in the coal industry:

> Mr. Kennedy has described coal as poison and those of us in the coal industry as poison dealers. That's a very serious accusation. He is telling our coal miners, our coal transporters, our coal power generators—and their families— that they're accessories to murder.
>
> *Nothing* could be further from the truth.
>
> To say something is poison means that it makes you very sick or kills you. But when countries generate electricity using coal, they lead healthier and longer lives.

In the last 20 years, countries such as China and India have started using many times more coal, and their health and longevity have shot up. They are buying it voluntarily because it is good for their lives.

It's estimated that, in large part thanks to new, coal-powered infrastructure, between 1 billion and 2 billion people now have access to clean drinking water that didn't 20 years ago.[24]

Do you know what clean drinking water means to a child who can play with his friends because he's not deathly ill with some parasite? Do you know what this means to a mother who doesn't have to worry about the water she gives her child, morning, noon, and night?

Without coal, countless children would be unnecessarily sick. Is Mr. Kennedy saying we should turn back the clock? There are still nearly a billion more people without clean drinking water, whom coal could help.[25] Is Mr. Kennedy saying we shouldn't go forward? Coal is the opposite of poison—it is medicine.

Now coal has certain risks—as does medicine.

Coal's risks come from the fact that historically it was formed from super-compressed ancient plants.

As a result, coal contains natural plant elements like nitrogen and sulfur, which are benign in modest quantities but harmful in larger quantities.

Therefore, it's important to limit the amount of these materials that come out of coal plants near large population areas—which is exactly what we in the American coal industry do. And that's what we encourage other countries to do.

If Mr. Kennedy truly cares about human health around the world, he should join the coal industry in the campaign to free coal exports while calling for better pollution laws abroad."

What is Kennedy going to say to this? What's any anti-coal person going to say about this? In my experience, it's hard to say much.

CHAPTER 5: HOW THE "GREEN ENERGY" MOVEMENT HARMS THE PLANET

Imagine a medical activist group introduced a bill to ban antibiotics — even though antibiotics have eradicated dozens of deadly diseases. Why ban antibiotics? Because, the group says, antibiotics are "dirty remedies". Antibiotics have all kinds of problems, they say, and some of the problems they cite are true. For example: Antibiotics can make diseases stronger and more resilient. They can cause serious allergic reactions and side effects. And so on.

But what about the benefits of antibiotics? Don't worry, the group says, antibiotics can be replaced with better, "green" remedies.

Would we buy this argument and ban antibiotics? Of course not. We would recognize that it is easy and common for people to attack good things by just focusing on their challenges, and by inventing harms. It's easy to exaggerate problems instead of trying to solve them. And it's incredibly easy to make up a "superior solution"...in your imagination.

We would wish "green remedies" luck in proving their case to consumers. But under no circumstances would we allow out-of-context criticisms and empty promises to justify banning something that, in the full context, is so crucial to life.

At least we wouldn't in medicine. In energy, my field, I believe that the "green energy" movement is demanding the equivalent of banning antibiotics.

It is claiming, without one supporting example, that "green energy," mostly solar and wind, can power entire societies. It is demonizing as "dirty" the only three proven, practical sources of industrial-scale energy — fossil fuels, nuclear, even hydroelectric.

It uses any attack it can find against so-called "dirty energy."

Some attacks are based on genuine concerns, such as: without modern filtration systems, coal plants emit particulate matter that, in sufficient concentration (but not in low concentrations), can negatively affect health. [26]

Many attacks are outright falsehoods, such as: nuclear power plants cause cancer and genetic mutations. [27]

And many attacks have no clear connection to health or economics, such as: hydroelectric dams should be stopped for the sake of "free-flowing rivers." [28]

On the basis of these attacks, which focus on all negatives and no positives, they are fighting for total or partial bans on fossil fuels, nuclear, and hydro.

Take Al Gore, a leader of the "green energy" movement. In a landmark 2008 speech, Gore claimed, without evidence, that he knew how "green energy" could give us the equivalent of $1 per gallon gasoline. [29]

But instead of starting a business to make billions on what would be the greatest energy breakthrough in decades, Gore proceeded to call for a ban on all fossil fuel electricity by 2018.

In fact, he implicitly called for an eventual ban on virtually all forms of energy. The only forms of power generation Gore supported as truly "green" were wind, solar, and geothermal energy—which today produce an unreliable and extremely expensive 2% of our nation's electricity that could not exist without fossil fuel and nuclear backup for when the sun doesn't shine and the wind doesn't blow. [30]

Gore's plan, and other plans that resemble it—such as Barack Obama's plan to cap fossil fuels 80% over the next several decades —are, taken literally, beyond catastrophic and are unlikely to pass. [31]

But it's important to understand that to whatever extent these ideas make it into law, it will be disastrous for both our economy and our environment.

If this seems implausible, it's because we have been taught to take for granted the amazing, unprecedented economy and environment we enjoy—and we have not been taught what they depend on.

Our lives are absolutely amazing. No one in history could have imagined a society in which the average person lived to 80, and, even more importantly, where there was so much vitality possible in those 80 years.[32]

Look at what we have access to: endless fresh food and clean water; racks of clothing; climate-controlled, weatherproof shelter; cures for practically every ailment; state-of-the-art hospitals; transportation anywhere in the world; thousands of career choices; boundless learning opportunities; endless social options; hour after hour of leisure time; and a lifetime worth of exciting things to do during that leisure time.

Because this is what we know, it seems guaranteed. But it's not. It can be lost. Because it is completely and utterly dependent on our ability to produce energy.

Our energy is our capacity to do work—from building a hospital to powering a hospital to transporting a hospital's daily supplies. The more work we can do, the more productive we can be, the longer and more happily we can live. Conversely, the less work we can do, the shorter and less happily we can live. The reason that the historical life expectancy is 30—is that human muscles and animal muscles aren't sufficient to do the work necessary for a high standard of living, including the immense amount of *time* our standard of living provides for scientific and technological research.[33]

The breakthrough of all breakthroughs was the development of *industrial energy*—energy used to power machines that could do superhuman amounts of work. This is not an easy task, as evidenced by the fact that we have only three forms of industrial-scale power generation: hydro-electric, which harnesses the power of large amounts of downward-flowing water (though such locations are limited); nuclear, which uses the power released from splitting atoms, and fossil fuels, which use the power of concentrated plant matter.

One of the most underappreciated benefits of industrial energy is the improvements in our environment it has produced. We're taught to think of our environment as something that starts out healthy and then we make dirty. The opposite is true. Nature does not give us a healthy, sanitary environment to live in. Try keeping your home environment clean without "unnatural" indoor plumbing, sewer systems, and garbage collection.

To live a truly human life, we need to radically *transform* nature using industrial-scale energy to create a truly *human* environment.

Can solar and wind do all these things? At this point, not even close. For sure, there's an enormous amount of energy in sunlight and in wind. But that energy is not concentrated in any single place—so it takes a lot of land and resources to collect. Worse, the energy doesn't come in as a reliable flow, it comes in on-and-off, intermittently. There are all sorts of fantasy schemes for making intermittent energy reliable, but none have solved this problem. That's why more solar panels in Germany have been coupled with more coal plants. [34]

Our response to "green energy" hype should be the same as our response would be to "green antibiotics"; if you're idea's so good, then prove it—but keep your hands off my livelihood. Billions of lives hang in the balance with energy production. In the past two decades, hundreds of millions of people have risen out of poverty because electricity production, overwhelmingly from coal, has skyrocketed—for example, it has quintupled in China.[35] 2 billion people have clean drinking water who didn't 20 years ago.[36] These gains, and the many before them, will disappear if the practical energy underlying them disappears. If we pass "green energy," we will have blood on our hands.

"Green energy" has nothing to do with protecting the human environment. The way to do that is through real energy along with clear laws protecting individuals and their property from pollution and endangerment.

But such laws must be rational; they cannot demand that energy production generate no waste or carry no risk, because waste and risk are inherent in life.

At any stage of development, the energy needed to improve our environment is going to have waste products. Some of these, such as particulate matter from coal, in isolation and in large quantities (such as in cities in the 1800s) have negative health impacts even though the technology overall is a boon to human health.

And over time, prosperous individuals have more time and technology to create progressively cleaner ways of generating energy. In generations past, people had to inhale large amounts of coal smoke in their homes, because it was either that or live an even more noxious life on the farm. Thankfully, that's no longer necessary; today's coal is hundreds of times cleaner than coals of generations past.

What about climate change? As in so many issues, we need to look at the full context. Now, part of that context, future predictions of how the greenhouse effect interacts with various hypothetical feedback loops, is very uncertain.

There is one certainty that is almost never mentioned, though. The livability of a climate, any climate, depends on its amount of industrial energy.

Consider the last 80 years. We hear about an allegedly alarming one degree Celsius temperature increase, yet who knows that the number of weather and climate related deaths decreased by *98%*?[37] If you care about the climate's effect on humans, you must embrace fossil fuels—and you certainly must demand that our legal system take a rational attitude toward clean, safe, non-CO_2-emitting nuclear power.

The uranium inside a nuclear power plant can't explode. Not one individual has died from radiation due to a modern nuclear power plant—not at Fukushima[38], not at Three Mile Island.[39] The majority of the people that died as a result of Chernobyl were involved in the rescue following the accident, which was caused by a combination of an unstable design and operator error.[40] The energy in nuclear material is a million times more concentrated than the energy in materials used in other forms of energy production.[41] That means it has exciting economic potential, if it weren't so controlled by government. And it means that the waste it generates is small, and, as France has shown, easy to deal with.[42]

But the progress of this safest of energies has been stopped by decades by the "green energy" movement, which claims that nuclear energy is going to turn us into cancer patients or mutants despite decades of Nobel-Prize winning physicists trying to educate them.

Which raises the question: Why are advocates of "green energy" so hostile to the technologies that make a truly livable environment possible?

CHAPTER 6: UNDERSTANDING THE GREEN MOVEMENT

What does "green" really mean? It is most commonly associated with a lack of pollution and other environmental health hazards, but this is both far too narrow and highly misleading. Consider the range of actions that fall under the banner of "green." As industrialists experience, it is considered "green" to object to crucial industrial projects, from power plants to dams to apartment complexes, on the grounds that some plant or animal will be impacted—plants and animals that take precedence over the *human* animals who need or want the projects.

It is considered "green" to oppose not only fossil fuel plants (which produce 87% of the world's energy), but hydroelectric plants and nuclear plants—which all told means 98% of the world's energy production.[43] It is considered "green" to turn off the heat or air-conditioning, even at the price of personal discomfort.

It is considered "green" to do less of anything industrial—from driving to flying to using a washing machine to using disposable diapers to consuming pretty much any modern product (there is now an attack on iPhones for being insufficiently "green" given the various materials that must be mined to make them).[44]

Often the same activity will be characterized as both "green" and non-"green"—just ask the proponents and opponents of any given solar farm. The proponents will say that the installation is "green" because it doesn't use fossil fuels (except, they evade, to mine, fabricate, transport and assemble it and to run and maintain the finished installation), it isn't mining the earth's precious "natural resources" (except, they evade, for enormous amounts of steel, aluminum, concrete, and various rare, and toxic elements), etc.[45] The opponents will point to the fact that solar farms, because they use a diffuse, intermittent energy source, take up an enormous "footprint" on nature through land use, that they require prominent, long-distance transmission lines to take to their customers, that they require large-"footprint" backup systems to store energy or fossil fuel plants to serve as backups, etc.

Clearly, "going green" is not primarily about human health—indeed, in its opposition to just about anything industrial, it threatens the industrial foundations of modern health and sanitation. The essence of "going green," the common denominator in all its various iterations, is the belief that *humans should minimize their impact on nature*—that the transformation of nature is immoral.

"Green" leaders and followers may disagree on how to implement this ideal, and they certainly do not follow it consistently, but nevertheless it is uncontroversial that minimizing impact is the ideal.

But if we take ideas seriously, then the "green" ideal should be more than controversial. It should be jettisoned, as it is squarely opposed to the requirements of human life, including the requirements of a *healthy human environment*.

Human beings survive by transforming nature to meet our needs. The higher our level of survival—the longer we live and the fewer of us avoid an early death—the more we must transform nature. In other words, we survive to the extent we depart from the "green" ideal.

Nature does not provide us with the wealth or the environment we need to live long, healthy, happy lives; hence the historical life expectancy of 30.[46] To live and thrive, we must create wealth and create a livable environment. And every new act of creation, from building a fire to building an air-conditioned home to building the Internet, constitutes additional impacting—transforming—nature.

The fundamental reason for today's incredibly high standard of living is that thanks to industrialization—the pervasive use of man-made power to fuel industrial machines—human beings can do hundreds of times more work to transform nature than we could even 200 years ago. But if our ancestors had followed "green" strictures, industrialization would have never got off the ground.

When the early oil industry turned night into day by making cheap illumination available to millions, they did it by drilling thousands of deep holes in rural Pennsylvania, extracting the black gold beneath, refining it into various useful substances, burning kerosene to create light, and dealing with whatever waste products emerged. J. J. Hill's Great Northern Railway, a private transcontinental railroad that revolutionized American transportation and commerce, required men to mine iron ore from the ground, to combine it with carbon to make steel, to mine and use coal to power the steel furnace, to pour the mixture into molds, to use the molds to make railroad tracks, to lay the railroad tracks across patches of wilderness, to displace various plants and animals that stood in the way, and many more changes to the status quo.

Fast forwarding to today, the Chinese airports and buildings that many marvel at also transform nature on a massive scale—from the magnitude of the physical structures themselves to the coal plants, gas plants, factories, mining operations, oil rigs, oil refineries, and heavy machinery that went into building them, not to mention the industrial transportation system that keeps them maintained and stocked with supplies.

Industrial progress is not "green." "Going industrial" requires a commitment to impacting nature as much as necessary to make it more hospitable to human life. And it is no accident that in generations past, Americans viewed industrial progress, not industrial abstention, as an ideal to strive for. Earlier generations took pride in transforming nature—in being a people that "tamed a continent," that built new factories, that paved new roads, that drilled new wells, that mined the earth for new resources. Whole towns would celebrate when a new bridge was built, when a factory was erected. They would proudly drive their automobiles, fly in planes, support new railroads, build new roads—without a shred of guilt over the fate of the snail darter. [47]

What about "green" support for "green energy" and a "green economy"? Is this not just a new, superior form of industry? Far from it. Any talk of green industry is ultimately contradictory, which is why such industries never materialize on a significant scale. All energy production requires an enormous amount of industrial development, both in its production and in its consumption.

Thus, environmentalists frequently oppose every power source, including solar and wind, for their various impacts. (They complain that solar and wind farms have the largest land "footprint" of any form of energy generation, which is true. [48]) Similarly, for all the talk of "green construction," "green building," and "green jobs," any activity with a major industrial presence will draw "green" opposition—as the valuable website www.projectnoproject.com aptly details. [49]

The more consistent anti-industrialists are explicit about their goal, including its ultimate implication: de-development and depopulation. Stanford environmentalist celebrity Paul Ehrlich, who likens population growth to a "cancer":

> A massive campaign must be launched to de-develop the
> United States. De-development means bringing our economic
> system into line with the realities of ecology and the world
> resource situation. [50] [51]

Billionaire Ted Turner, a "mainstream" figure, says: "A total [world] population of 250-300 million people, a 95% decline from present levels, would be ideal." [52]

The true nature of "green" emerged particularly clearly in a debate over nuclear fusion in the late 1980s. Some uninformed news reports announced that fusion—which, if it worked, would be the cheapest, cleanest, most plentiful source of energy every created—was on its way to commercial reality. Many expected environmentalists to embrace this development. They condemned it.

"It's the worst thing that could happen to our planet," said leading environmentalist Jeremy Rifkin. Ehrlich memorably said that allowing human beings to use fusion was "like giving a machine gun to an idiot child."[53] Environmentalist icon Amory Lovins stresses he would oppose any fusion-like energy breakthrough: "Complex technology of any sort is an assault on human dignity. It would be little short of disastrous for us to discover a source of clean, cheap, abundant energy, because of what we might do with it."[54]

Do not make the mistake of writing off these anti-industrialists as "extremists" who don't reflect on "moderate" greens. While the "extremists" are more *consistent* than the "moderates," they share the same ideal—the anti-impact ideal that destroys industrial progress to whatever extent it is practiced.

But what about the "environmental impact" of industrial development? Isn't the "green" movement providing a salutary influence to us by helping us combat that problem? Again, no.

The idea of "environmental impact" is what philosopher Ayn Rand called an "intellectual package-deal." Such a concept dishonestly packages together two very different things—the impact of development on the *human* environment and the impact of development on the *non-human* environment. Industrial development will certainly often harm various non-human environments—but it is a godsend to the human environment. By lumping together concern with the non-human environment (e.g., displacing some caribou to get billions of barrels of the lifeblood of civilization) and the human environment (e.g., air quality), anti-industrialists are able to dupe Americans into thinking that sacrificing to caribou somehow benefits them.[55]

Historically, industrial progress brought with it a radical improvement of the human environment. Indeed, industrial progress essentially *is* the improvement of the human environment. The reason we develop is to make our surroundings better so that our lives are better, cleaner, healthier, safer—in the face of a natural environment that is often hostile to human life.

Contrary to "green" mythology, man's natural environment is neither clean nor safe. In a non-industrialized, "natural" state, men face all sorts of health dangers in the air and water, from the choking smoke of an open fire to the feces-infested local brook that he must share with animals. Industrial development gives men the technology and tools to make their environment healthier—from sanitation systems to sturdier buildings to less onerous job conditions to comfortable furniture to having healthy, fresh food at one's disposal year round, to the wealth and ability to preserve and travel to the most beautiful parts of nature. And so long as we embrace policies that protect property rights, including air and water rights based on sound science, we protect industrial development and protect individuals from pollution.

As for the "unsustainability" of industrial progress, an accusation that dates back to Marx, this fails to recognize the fact (elaborated on by Julian Simon and Ayn Rand) that man has an unlimited capacity to rearrange nature's endless stockpile of raw materials into useful resources and gain more resource-generating knowledge as he proceeds—which is why the more resources we use, the more resources we have.

Human life requires changing nature on a massive scale. Any cause that holds minimal impact as an ideal is anti-human and an enemy of the human environment.

Today's anti-industrial movement is not new in this respect. Throughout history, there have been major anti-industrial groups or movements. The basic premise they have in common is that it is arrogant and wrong for man to transform nature as he sees fit. Man, they believe, should not tame nature but exist in some sort of mystical "harmony" with it (how he is supposed to cope with nature's dangers and a life expectancy of 30 is rarely specified). Perhaps the most iconic anti-industrialist was the 18th Century's Jean-Jacques Rousseau, who worshiped nature untouched by man and regarded the transformation of nature in his time (let alone the then-unimaginable transformation that is our modern world) as evil.[56]

But the modern-day followers of Rousseau knew they could not succeed by being directly anti-industrial. So they create a false association between themselves and environmental progress, and a false opposition between industrial progress and environmental progress.

Part of this false conceptualization has been achieved by using an old socialist trick to obscure the massive environmental improvement that industrial capitalism brought. The trick is to criticize something *by comparison to a nonexistent and impossible utopia.*

Socialists used this technique to criticize capitalism for causing poverty, even though capitalism *inherited* poverty—and cured it. Yet Marxists would attack capitalism's incredible contribution to human life, including to the life of laborers, by comparing that contribution, not to its predecessors and not to any known alternatives, but to a fictional socialist utopia whose advertised results contradicted everything known (even then) about socialism's destructive nature.

Environmentalists have done the equivalent to industrial progress. Instead of comparing the human environment pre-industrial and post-industrial, they *compared the post-industrial environment to a non-existent pollution-free utopia* achieved by man living in "harmony" with nature. They did this in spite of conclusive historical evidence living in "harmony" with nature means living very briefly. Historically, to the extent humans didn't mine, didn't burn fuels, didn't develop, and were unwilling or unable to control or displace other species where necessary, they died early and often. The modern standard of living is an unprecedented, singular achievement that continues only so long as men are free to command nature on a large scale.

Early environmentalists cursed the coal fumes of newly industrial cities, evading the wood fumes, dung fumes, and starvation coal had replaced—and the work-hours it saved and years of life it added to human life. They cursed smog, evading that it replaced rampant airborne disease from horse-drawn society.[57] And when increased production of coal and oil and natural gas produced the energy and technology to develop ways to radically reduce their pollution, environmentalists took credit—as if laws against pollution weren't essential to capitalism, the system where protection of all forms of property is sacrosanct.

Development, industrial progress, and capitalism promote a *human environment*. The anti-industrial "green" movement opposes it. This is a truth that Americans desperately need to understand. At present, the philosophical confusion caused by anti-industrialists causes Americans who are genuinely concerned about their health and well-being to embrace the ideas and policies of those who want to sacrifice that health and well-being to the non-human. We are taught to denigrate fossil fuels, without which most of us would have already died, and to strive for a mythical "green energy" economy, powered by fuel sources that have failed for decades. We are not taught that industrialization has enabled man to be orders of magnitude less vulnerable to climate, but that, contrary to all climatological history, a degree Celsius rise in temperature over 150 years portends catastrophe. With proposals on the table such as

80% cuts in CO_2 emissions, "green" confusion could mean economic suicide.

Such is the power of moral idealism and philosophical corruption. The ideal—and the corruption—need to be replaced.

CHAPTER 7: BILL MCKIBBEN—ENERGY ENEMY NUMBER ONE

Bill McKibben, who has been called "the nation's leading environmentalist," is leading a movement to destroy the fossil fuel industry, which he calls "Public Enemy Number One." This is the signature issue of his mega-popular organization 350.org, which orchestrates campaigns such as "Do the Math" and "Fossil Free."

As an energy researcher who knows the indispensability of the fossil fuel industry to my own life and billions of lives around the world, I am doing whatever I can to stop this movement.

My Debate with Bill McKibben

I publicly debated Bill McKibben in November 2012 in order to make the case that his quest "to cut our fossil fuel use by a factor of 20 over the next few decades" is pseudoscientific and suicidal.[58]

Throughout the debate I stressed four points:

1. For the foreseeable future, fossil fuels are *the* indispensable source of the abundant, affordable energy that human flourishing depends on.

2. The proven science about climate illustrates a mere half-degree warming in the last 70 years, including virtually no warming in the last 15—McKibben's claims of catastrophe are based on the extreme speculation of climate prediction models that can't predict the climate.[59]

3. The overall impact of fossil fuel use and the technologies it powers has been to make our climate dramatically safer—climate-related deaths have fallen 98% since 1920.[60]

4. The world desperately needs more energy—2 times as much if everyone is to get to the same level as Germany—and yet McKibben is calling for 95% of fossil fuels to be illegal.[61]

Readers should watch the debate and draw their own conclusions, but from my vantage point the thing that struck me most about McKibben's approach was that he was intellectually and emotionally indifferent to the fundamental importance of affordable, abundant energy.

Sloppy Thinking: From Organic Farming to Solar and Wind

For example, in agriculture, where oil and natural gas are the difference between abundant food and mass-starvation, I said:

> If Bill McKibben came here tonight…and said, my conclusion is we should ban 95% of food, you would say that's crazy. But he is saying we should ban 95% of fossil fuels—which is the food of food. Without fossil fuels, billions of people will starve. There is no evidence to the contrary and so to cavalierly talk about that is just…really, really irresponsible because these are real lives. These are people who if we do the wrong thing, they will die and ultimately you know you will suffer too but these people will die and one thing we know is that modern industrial fossil fueled agriculture saves billions of lives. And what Bill is saying would take them away.[62]

McKibben's response was to cite a single paper declaring that in one region "organic yields were essentially equal to this point to yields from conventional farming."[63]

One thing striking about this was his willingness to equate one cherry-picked paper with an objective big-picture analysis—when the stakes are the very survival of billions of people.

But another, even cruder error, is to ignore that "organic" agriculture uses immense amounts of fossil fuels. Unless McKibben has a study to show that men with shovels can equal the production of men with tractors.

This pattern repeated itself time after time—McKibben would rationalize his radical, ruinous prescription with offhand sloppiness.

He exhibited the same sloppiness when discussing the alleged replacements for fossil fuels: solar power and wind power. Since McKibben uses Germany as an exemplar of a "green energy" future, I pointed out that Germany has not replaced a single coal plant with solar and is building over a dozen new coal plants—because solar and wind are nowhere solving their intractable problems of unreliability.[64]

McKibben responded that the unreliability problem, which has rendered solar and wind energy failures for more than 75 years, was no longer a problem—offering as evidence a story he read in that morning's news:

> Renewables really work. There is nothing speculative anymore about them. In fact, and again this is why it's important to listen to dates and to evidence, there's a report this morning from the German ministry, energy minister, Stephan Kohler, who works of course in the conservative government of Angela Merkel that the country will easily beat even its own ambitious plans for renewable energy and generate more than half the country's power that way by 2025 and perhaps as high as two thirds.[65]

Let's leave aside the fact that, whatever McKibben read in the morning paper, Germany has no energy ministry and therefore no energy minister. McKibben is completely misrepresenting Kohler; see this recent interview with Kohler in which he says: "They say that we could replace power plants operated with fossil fuels by adding more renewable energy sources. My response to them is: It won't work."[66]

Further, McKibben, knowingly or not, is regurgitating what amounts to energy accounting fraud. The German government and others cannot replace reliable coal plants with unreliable solar panels and windmills, but to garner international praise they inflate their numbers by pretending that the sun shines 24 hours a day and the wind blows 24 hours a day.[67]

If Bill McKibben were engaging in pseudo-journalism and pseudo-science and pseudo-economics on some obscure blog, I wouldn't care. But, as I reminded him more than half a dozen times during the debate, he is an intellectual superstar using his enormous platform to call for 95% of our most important source of energy to be outlawed, which, on the basis of everything we know, would ruin billions of lives. Every time I raised McKibben's stated goal, he dodged the issue, at most voicing empty platitudes such as "I never said it would be easy."[68]

Tell that to the ambitious young Chinese man who, had nations followed McKibben's past guidance, would have never gotten his first light bulb, his first refrigerator, his first decent-paying job.

Tell that to the Indian mother whose child would have died of starvation were it not for that country's fossil-fuel-powered agricultural revolution.

Enemy of Energy

Bill McKibben is Energy Enemy Number One. And he's particularly dangerous because he is taking the moral high ground against fossil fuels, which is the most powerful rhetorical position to have. But he does not deserve that high ground, and we who value affordable, abundant energy need to take it away from him.

The 350.org movement to morally condemn the fossil fuel industry needs to be outmatched by a movement to morally *champion* the fossil fuel industry and the energy industry more broadly (McKibben opposes the vast majority of nuclear and hydro). We at the Center for Industrial Progress are starting such a movement.

CHAPTER 8: THE "HIDDEN COSTS" ("EXTERNALITIES") FALLACY

Opponents of fossil fuels have long championed solar power and wind power as replacements. Unfortunately, there is no evidence that solar and wind can provide the cheap, plentiful, reliable energy that our standard of living requires. They have never come remotely close to competing economically on a free market. In fact, due to their low concentration and high intermittency, they have proven unable to provide substantial baseload power in any country, ever, even when exorbitantly subsidized. Thus, they do not qualify as "energy" in the modern sense.[69]

When confronted with these facts, opponents of fossil fuels offer a seemingly scientific counter-argument. Fossil fuels are only cheap, they say, because fossil fuel companies aren't required to pay for the "hidden costs" or "negative externalities" of their product.[70] These "hidden costs" are harms not reflected in the prices we pay—such as the presumed damage from future climate change. Companies should be required to pay these "hidden costs," the argument goes, and if they were, solar and wind would actually be cheaper than fossil fuels.

In a recent column, "Here Comes the Sun," Paul Krugman invokes this view to argue for major taxation on fracing (and, by implication, all other fossil fuel production). To believe otherwise, he says, is to be economically illiterate.[71]

> Fracing—injecting high-pressure fluid into rocks deep underground, inducing the release of fossil fuels—is an impressive technology. But it's also a technology that imposes large costs on the public...Economics 101 tells us that an industry imposing large costs on third parties should be required to "internalize" those costs—that is, to pay for the

damage it inflicts, treating that damage as a cost of
production.

Unfortunately, this analysis fails both Political Philosophy 101 and,
surprisingly given Krugman's credentials, Economics 101.

It is true, as Krugman says, that the price of a product does not
reflect all the negative impacts that come with the product. For
example, when the automobile industry overtook the horse-and-
buggy industry, there were many negative impacts on the workers
and families of the latter industry, who had to suffer temporary
unemployment, go through retraining, etc.

But Economics 101 does not tell us what to do about such effects.
For example, it doesn't say whether the automobile industry should
have been forced to pay a tax for "imposing large costs on third
parties" it drove out of business. Such questions are the province of
political philosophy. Krugman is welcome to argue for his personal
political philosophy, which, in my reading, is a hybrid of utilitarianism,
egalitarianism, and economic authoritarianism. But he should not
abuse his economic prestige to smuggle his political views under
"Economics 101." The idea of dealing with pollution issues via
"externality" calculations, rather than by proper definition of air and
water rights, is highly dubious and anything but self-evident.

That said, it can be valuable in understanding the economic impact
of an industry to analyze negative impacts that are not reflected in
prices. But we must simultaneously analyze the *positive* impacts that
are not reflected in the prices we pay. But Krugman and others
steadfastly refuse to consider the "hidden benefits" of fossil fuels—
even though they are massive.

A very clever video on YouTube illustrates the issue of hidden
benefits with regard to the Internet. "How much money would
someone have to pay you," the host asks, "to give up the Internet for

the rest of your life?"[72]   In other words, how much is the Internet really worth to you? The video's featured economist, Professor Michael Cox, says his students mostly answer that no amount would be enough—and when they propose amounts, they are in the high millions or billions. Most of us would say the same, because the Internet is an indispensable value in our lives.

And yet how much do we actually pay for it? Less than a thousand dollars a year. "What the market has done," observes Cox, "is create a tremendous gap between worth and cost."[73] This gap is a wonderful thing—so long as we don't forget it when assessing the importance of indispensable values to our lives.

An equivalent gap between worth and cost exists between what we get and what we pay for indispensable sources of cheap, plentiful, reliable energy such as coal, oil, and natural gas—since energy is the resource that makes every other resource in our industrial economy possible.

Consider: If you were a factory owner, how much more would you be willing to pay for the coal-powered electricity that allows your business to exist? How much would you be willing to pay for the natural gas that keeps you from freezing in the winter? If you are a parent, how much would you be willing to pay for the gasoline in the ambulance that saves your child's life? A lot more than you do. The reason we get energy for such a bargain is because of the wondrous nature of the free market, including another part of Economics 101 Krugman conveniently omits: the marginal nature of prices. The price we all pay for a given good or service is set by the marginal buyer— the buyer who, among those having successfully bid for the good or service, was willing to bid the least for it.

This means that every other buyer valued the good more than the price paid. Thus, with every product or service, the total value consumers gain from buying it is necessarily higher than the total price they pay for it. And in the case of indispensable values, such as the Internet or cheap, plentiful, reliable energy, that value is incomparably higher. We should never confuse the price we pay for fossil fuels with the value we get from fossil fuels.

An honest attempt to guesstimate the full economic impact of fossil fuels would have to take into account, at the absolute minimum, the following:

1. Cheap, plentiful, reliable energy adds dozens of years to life and makes those years incomparably more enjoyable.

2. Given current technology and economics, including the desperate need for industrial-scale energy around the world, fossil fuels will be indispensable for decades to come. This is especially true because of the environmentalist assault on nuclear power, which has set back that technology decades.

3. Industrial-scale energy has historically made us far less vulnerable to climate, not more—and would be essential to coping successfully with any serious climate change, natural or manmade.

4. Solar and wind have never produced cheap, plentiful, reliable energy in any country, ever.

None of this enters into Krugman's "scientific" evaluation. He treats the price of fossil fuels as fully reflective of their positives, and regards it as scientific to fixate on their negatives (real or fabricated) and demand that massive taxes be levied. What level of taxation? Krugman doesn't say—but let's explore the alternatives.

If the Left imposed a carbon tax that was large enough to force the entire economy to run on solar and wind, the entire economy would collapse. If the tax was large but not large enough to totally bankrupt the fossil fuel industry, it would do little to reduce greenhouse gases but make us far poorer, including far more vulnerable to the climate—cheap energy being the key to making the climate livable. Any level of tax is pseudo-scientific and destructive, because it is based on an evasion of the indispensable, life-and-death positives of fossil fuels.

What should the government's policy toward pollution be? This is a complex subject, but in my view the proper principle to guide policy is individual rights. The government should clearly define air and water rights, and enforce individual cases according to objective evidence of physical harm.

How such rights should be defined is in part an issue of the economic and technological context. Since it is never possible to eliminate all challenges pertaining to waste, what if and when a given amount of waste constitutes a rights violation depends on the full context of what is preventable in a given economic and technological context and what isn't. In today's context, to call CO_2 emissions "pollution" is to call human survival "pollution." No view could be more damaging to our economy—or to the human environment.

CHAPTER 9: THE "RENEWABLE ENERGY" FALLACY

The notion of "renewable energy" has two fundamental conceptual flaws. It's not really renewable, and it's not really energy.

"Renewable" in most definitions approximates to something like "naturally replenished" and it often contrasted with allegedly inferior, "finite" sources. It brings to mind the image of a pizza where a slice, once eaten, magically reappears.

There is no such phenomenon in nature, though. Everything is finite. The sun and the photons and air currents it generates are not infinite; they are just all part of a very large nuclear fusion reaction. True, that nuclear fusion reaction will last billions of years, but so will the staggering amounts of untapped energy stored in every atom of our "finite" planet.

To obsess about whether a given potential energy source will last hundreds of years or billions of years is to neglect the key issue that matters to human life here and now: whether it can actually provide the usable energy that will maximize the quantity and quality of human life now. Even if there was an "infinite" energy source, if it was worse than a superior source that would last 50 years, we should of course use the superior source first—and work toward discovering an even better one. An infinite life of inferiority is hardly ideal.

This is borne out by the history of energy production. For most of human history, our amount of usable energy was barely above the amount needed to power our muscles (and during famines, not even that). There was copious amounts of unusable energy—the chemical bonds in deposits of coal, oil, and natural gas, the mechanical energy of the wind, the photons of the sun, and, greatest of all, the energy stored in all the matter around us, whose proportions were quantified when Einstein identified that $E=MC^2$.

Every advancement in energy production consisted of taking some unusable source of energy and rendering it usable—windmills for grinding grain, water-wheels for operating simple machines, and ultimately, concentrated hydrocarbon fuels that multiplied human productivity hundreds of times over.

Hydrocarbons et al are often called "finite natural resources," but this is a misnomer; they are not naturally a resource. They become resources—i.e., they deliver services—only insofar as they are rendered valuable by human intelligence. This is Julian Simon's crucial identification that the human mind is "the ultimate resource" that creates new resources, including energy resources, by discovering how to extract new services out of previously useless raw materials. We should not think of unusable raw materials as resources until or unless they are rendered usable by human intelligence.

This last applies to the sun (and the wind), the ultimate source of most "renewable" energy. The vast majority of sunlight does not provide usable energy given any known technology. True, through photovoltaic conversion, a solar panel in most places can generate an electrical current of some magnitude. But who cares? A hurricane produces many h-bombs worth of mechanical energy—does that make it an energy resource?[74] Not if it can't be harnessed in a manner that provides the cheap, reliable power that we can use to meet our present and future needs. In the vast majority of cases, solar conversion technology can't, the energy collected is too dilute and intermittent to be a useful source of large-scale energy. (And if it could be, imagine the "environmentalist" opposition to the amount of space the panels took up and the amount of industrialization we performed with the energy!)

So "renewable energy" as it is commonly used to mean solar and wind, is not "energy" in the economic sense of the word. It is a hypothetical source of energy that we know of, but that hypothetical deserves no more privileged status than any other kind of hypothetical (the ability to unleash atomic energy from a wide range of elements) let alone methods with far more promising potential (e.g., the potential of uranium and thorium to generate tens of thousands of years worth of energy).

The idol of "renewable" energy is part of the broader idol of "sustainability." Both of these are false idols that obscure the true beauty of capitalism, which is that in producing energy—and everything else—it is better than "sustainable"—it is progressive. "Renewable" or "sustainable" implies that the ideal life trajectory is one of repetition, using the same methods and materials over and over.

But that is an ideal fit for an animal, not a human being. The human mode of existence is to always get better, always improve, always discover how to use new raw materials to create energy.

The root of the fetish with "renewable" energy is the Green ideal of minimizing man's impact on nature. This is borne out by the fact that the only practical "renewable" source of energy, hydroelectric, is widely opposed by the Green movement for interfering with "free-flowing rivers." That movement prizes solar and wind despite their horrendous track record for ideological, ultimately religious reasons: the idea of a society only relying on the sun and the wind is congenial to their ideal of a world in which man tiptoes on the planet instead of transforming it.

If we cast aside the Green religion, "renewable energy" is a false ideal that has no place in a rational discussion of energy. The only question that matters about energy is: What sources of energy will best advance human life now and in the relevant future (not 5 billion years)?

The only way to answer that question is to leave producers and consumers free to seek out ever-better answers in a free market. Then, we will always have the best kind of energy—progressive energy.

CHAMPIONING FOSSIL FUELS

CHAPTER 10: TAKING THE MORAL HIGH GROUND

Imagine you are an advertising executive, and a CEO asks you: "Do you think you can help improve the reputation of my industry?"

You respond, "Sure, what are some ways your industry makes people's lives better?"

He replies, "Well, actually, our product helps people in just about everything they do. This past year, it helped take 4 million newlyweds to their dream destinations for their honeymoons. It helped bring 300 million Americans to their favorite places: yoga studios, soccer games, friends' houses. It made possible the bulletproof vests that protect 500,000 policemen a year and the fire-resistant jackets that protect 1,000,000 firefighters a year." [75] [76]

"If you do all that, how could you be unpopular?"

"We're the oil industry."

Why is the oil industry so hated? After all, the oil industry does everything I said above, and many more wonderful things.

One common answer is that the oil industry has done bad things, such as the BP oil spill. (Though given that the Gulf of Mexico naturally "spills" two Exxon Valdez's of oil annually, the reaction to the spill was wildly out of proportion. [77]) But every decent-sized industry is going to have companies who do bad things. Many solar and wind companies, for example, shave costs on their expensive, unreliable energy by using materials from deadly Chinese rare-earth mines, and yet their reputation is outstanding. [78] Yet with oil, people see only negatives and no positives.

Before you blame the biases of the public school system and the media (which deserve plenty of blame) ask yourself this: How much do you hear from the *oil companies themselves* about all virtues of oil and oil production? Consider this. On the homepages of the three most prominent oil companies—ExxonMobil[79], Shell[80], and Chevron[81]—there is not one single mention of the word "oil."

These companies are obviously not comfortable publicly touting the virtues of their product. Why?

Because all of us, including oil companies, have been taught that the oil industry is not *moral*. We have been taught that there's something inherently wrong with transforming our world by drilling for oil and consuming it—whether to burn in an automobile or to make a plastic bag. We have been taught that in an ideal world, there would be no oil industry. The oil industry is, on this view, a necessary evil at best—and an unnecessary evil at worst.

The moral case against oil can be boiled down to two ideas:

1. The oil industry is inherently unsustainable. Using oil is short-range and self-destructive, and the oil industry is preventing us from adopting better, long-range solutions.

2. The oil industry is environmentally harmful. Using oil inherently pollutes the world around us, and we should use better, non-polluting technologies.

These ideas have become omnipresent—outside and inside of the oil industry. Ask yourself: "Do I believe the sustainability argument or the environmental argument? Do I think they're at least partially true?" Based on my experience talking to hundreds of people in the industry and observing thousands more, the answer is likely yes.

And that's why the oil industry is always seen negatively; its opponents use the moral objections against oil to take the moral high ground—and there is no more powerful position than the moral high ground.

But it is the oil industry, not its opponents, that deserves the moral high ground. The moral arguments against oil pretend to be progressive but are in fact re-hashes of primitive philosophical doctrines. For example, "sustainability" is a relic of centuries when human beings repeated the same lifestyle over and over—instead of finding better and better ways to do things.

The moral case against oil can be refuted and replaced by two concepts that marry energy knowledge and moral philosophy:

1. Progressive energy: The ideal source of energy is not some "sustainable"—i.e., endlessly repeatable—form, but the best, cheapest, ever-improving form human ingenuity can devise. As long as human beings are free, they will continue to develop new resources from previously useless raw materials (such as shale oil). An oil industry is ideal in the same way the iPhone is an ideal for so many. It may not be the best forever, but it is the best for now and we should be grateful to have it.

2. Environmental improvement: Energy and technology, including the oil industry, are needed to improve nature—which, left to its own devices, is resource-poor and threat-rich. Every activity has negative byproducts, but the net environmental impact of oil is a radical improvement.

Through these concepts and others, we can give the oil industry—and, more broadly, the entire energy industry—what it needs: a moral defense. This means an understanding, backed by 100% conviction, that the oil industry is fundamentally a force for good in human life. (If you want to see what this conviction looks like outside the oil industry, see the "I Love Fossil Fuels" campaign.[82])

This is why my organization teaches Energy Ethics 101 to the energy industry. The millions of people who work in this industry deserve to understand why what they do is right and that why those who try to take away their freedom are wrong.

CHAPTER 11: WHY WE SHOULD LOVE THE OIL COMPANIES

What should oil company executives do to improve their industry's reputation and secure their freedom to produce the lifeblood of civilization?

Unfortunately, the conventional answer is: pretend they're not oil companies. BP's John Browne some years ago infamously declared his company's aspirations to be "Beyond Petroleum"—a slogan that obviously does not aid the industry's desire for more petroleum drilling rights. (BP, to its credit, no longer trumpets this slogan, which defaults BP back to the implicit original, British Petroleum.)

Chevron's mega-PR-campaign, "We Agree," features 10 empty slogans, not one of which expresses pride in producing oil, and some of which are downright offensive. "Oil companies should think more like technology companies," the campaign says—as if the ability to extract the greatest portable fuel known to man from once-useless shale rock 10,000 feet below the surface of the Earth is not a technological achievement. [83]

This kind of posturing is self-defeating—no one believes that oil companies are anything other than oil companies. And it is a disservice to both their industry, which does not deserve flagellation (except when they rent-seek or engage in self-flagellation), and to the American people, who desperately need to know the positive importance of the oil industry in their lives.

We should never forget that the oil industry, whatever its problems (and most of those are caused by bad government policies) is the single most vital industry in the world.

It has revolutionized agriculture; without oil and natural gas-based agriculture, we would not have the fertilizers, tractors, and transport that enable farmers to feed a record 7 billion people with the lowest malnutrition level in history.[84] In other words, the oil industry, to exaggerate only slightly, "solved world hunger." Wouldn't that be profitable to point out?

The oil industry has revolutionized health care. Every hospital lives and dies based on just-in-time transportation of supplies, sanitary plastic devices and disposables, and petroleum-based pharmaceuticals. Without hydrocarbon-based synthetic pesticides, the U.S. would still be cursed with insect-borne diseases, such as malaria, which afflict much of the undeveloped world. Wouldn't that be profitable to point out?

I could multiply the examples to every other industry, because every other industry benefits in proportion to the availability of cheap, plentiful, reliable, portable fuel—and that is what the oil industry works everyday to bring to us.

The benefits of oil are all around us. If most Americans truly understood these benefits, they would surely have a different view of the industry. They would think more like 1920s best-selling author Bruce Barton, who said, "My friends, it is the juice of the fountain of eternal youth… It is health. It is comfort. It is success."[85]

As the Founder and the Director of the Center for Industrial Progress, I make it my job to educate the public about the incredibly positive role energy and industry, particularly the oil industry, play in their lives. For the last five years, I have been giving speeches around the country, especially at universities, about how the oil industry produces the lifeblood of civilization, and about how we should value the industry and above all value its freedom to produce.

You might expect that audiences would reject this message and write me off as an industry shill. But the exact opposite happens—because the truth is on my side and I don't hide it or apologize for it. I explain to them that I came to my conclusions after studying carefully the relationship between oil and human life over the past 150 years, and welcome them to do the same.

In fact, not only do audiences not run me out of the lecture halls, they get excited about oil production, and a little bit upset that they never learned this anywhere else. For example, most people are blown away when I point out how much of whatever room I'm lecturing in is made of oil—the insulation in the walls that kept us warm, the plastics in their electronics, the (synthetic) rubber in their shoes, the makeup on their faces, the glasses or contacts on their eyes, the paint on the walls, and so on. (And *everything* is transported using oil.) They're excited because this stuff is genuinely exciting, and because we are never taught it.

To be honest, I was initially surprised by how positive a reception I got: "After leaving his talk, I understood how rich my life is because of oil"; "Mr. Epstein's lecture made me realize that oil is a commodity which civilization cannot survive without and therefore its production is not only vital, but moral"; "I left with a greater appreciation of the role oil plays in my own life."

But then I realized why: by focusing on the positive of oil and the choice America faced about whether to pursue that positive to the next level, or forgo it and suffer, it made them care about and even love the oil industry.

I like to call this method of education "Aspirational Advocacy," because it means connecting our educational efforts with the audience's deepest values and aspirations. It is both the most genuine and most effective way I know of persuading people; I am not aware of any other approach that gets people outside the oil industry to love the oil industry.

America should love the oil companies, and if they change their strategy, millions more Americans will love the oil companies.

CHAPTER 12: TAKING THE ENVIRONMENTAL HIGH GROUND

Whenever possible in a debate, you want to take the high ground right out of the gate. When discussing fossil fuels, that is particularly true on environmental issues.

Here's an example of how to do it on coal, using the export terminals example from Chapter 3.

Here's what the industry might say to a college audience:

> There's a major new coal project in the Pacific Northwest that is a huge economic and environmental opportunity for America and our trading partners.
>
> Now it might sound odd to hear of coal as an environmental opportunity—but it's true.
>
> You may know that coal has dramatically improved the economies of India and China by allowing them to build super-productive factories that make their people much more well off financially.
>
> But you might not know that their environments have gotten much better as well.
>
> With cheap, plentiful, reliable energy, they have been able to better protect themselves from nature's dangers with things like:
>
> - water purification plants
> - irrigation systems
> - indoor plumbing
> - hospitals

- modern buildings

- disease control

- smoke-free electric stoves

Energy from coal has contributed to the increase in the number of people with clean drinking water by over 2 billion over the past twenty years.[86]

In the last 20 years India has multiplied its coal-powered electricity generation by more than 3 times, and in that time the average life expectancy has gone up by 7 years. [87] [88]

Now you might have seen or heard of harmful smoke clouds above China and other places that use a lot of coal.

But these exist not because they're using coal, it's because they're using coal improperly—without proper air and water-protection laws.

Coal, contrary to what you have heard, is not some scary, super-toxic material—is just super-compressed ancient plants.

Whenever you use plant-based substances to generate energy, you can run into problems with natural plant materials like nitrogen and sulfur—in certain quantities they're very healthy but in larger quantities they can be dangerous.

This is not just a coal risk—in fact, the worst air pollution in China comes from burning things like wood, straw, even animal dung in indoor or outdoor fires with no filtration systems whatsoever.

In many cases, the Chinese have highly-effective filtration systems for coal plants—the government just has companies keep them off.

We in the coal industry are encouraging China, India and others to pass proper air and water protection laws to protect citizens against all kinds of air-pollution.

That, combined with the incredible positive economic and environmental power of coal, will improve these countries' well-being greatly—and they will become even better trading partners.

A new coal mine is something we should all be excited about. And we have hit the motherlode of coal in a place in Wyoming called the Powder River Basin (PRB).

PRB coal is in huge demand in various Asian countries, who can use it to become more productive and raise their quality of life—which is good for everyone.

Are we going to take advantage of this enormous wealth creation opportunity—and international development opportunity—and leave industry free to build the necessary export terminals ASAP?

Or are we going to sacrifice prosperity and environmental health, here and around the world, to uninformed hysteria?

I hope you join our cause for the future of your community, your country, and the aspirations of billions around the world.

In my experience, this kind of argument is very hard to respond to. Because it owns the environmental high ground and covers all the bases, there's not much to attack—and a ton of positive claims they have to contend with.

One instructive experience I had in taking the environmental high ground was when delivering a provocatively titled talk "Why Mining Improves Our Environment" at an ultra-"green" campus.

Before the talk, a friend of mine noticed an anti-mining activist chatting on his cell phone saying how he was going to watch this crazy speech and then blast the speaker.

Given his past experience, he was expecting me to just evade the environmental issue, but I did the opposite.

I started the talk by expressing how important it was to be concerned about the environmental issues connected with mining.

I then read some excerpts about some horrible, toxic mining practices in China—the kind of thing you might read Greenpeace saying about coal, that always has the implication that if anyone, anywhere has a dangerous mine then of course we should ban coal.

And then I asked the audience, do you think we should ban the energy that is connected with these mining practices?

Many said yes.

And I said, by the way, this is a mine for the materials used to produce wind power, does that change your mind?

That threw them. And then, they took it all back and started making my points for me.

They say, come on, everything has negatives, but we need to look at the big picture, the overall impact. We should try to solve the problems, not throw out the baby with the bathwater.

I said I couldn't agree more, so lets look objectively at the environmental positives and negatives of mining in general and coal in particular. And then it was a slam dunk: mining improves our environment and coal improves our environment.

So in the question period, this guy and his friends who had planned to "gotcha" me were just sitting there silent and deflated. That was satisfying to me. Of course, I would have hoped that they became passionate supporters of the Center for Industrial Progress.

But if someone starts out as your seething enemy, making them deflated is a pretty good outcome, as those of you who deal with hostile regulators or activists would probably agree.

That said, the best outcome is making people inspired and into advocates.

CHAPTER 13: THE I LOVE FOSSIL FUELS CAMPAIGN

I've heard as an excuse in many industries that have to deal with the Green movement that we're at a disadvantage because the other side has some emotional advantage.

But that's only true if we let them own the value issues, like environment. If we own them, by giving the big picture, with plenty of examples, plenty of justified emotion—we have the advantage.

And in fact, people will be inoculated against anti-fossil fuel messaging, because they'll know clearly and concretely how destructive it is to oppose fossil fuel.

As evidence for this, I want to show you a few images from our new Facebook campaign, "I Love Fossil Fuels." I did not make one of these, they're all just from people who have taken in our work.

There's no reason this isn't possible on a bigger scale—all of us just need to work together to incorporate taking the moral and environmental high ground into our messaging.

We all have a stake in the war over fossil fuels, and it's a war that will ultimately be won or lost depending on whether we can win the environmental high ground.

Are you going to try to use this approach to help this industry earn the image, and the freedom that it deserves?

Or are you going to sit by and leave the fate of fossil fuels to others and to chance?

I hope you join us, because if enough of you do, and enough of your friends do, sooner or later a lot more people who used to say "fossil fuels are dirty" will say "fossil fuels are healthy"—or even better, fossil fuels are life.

APPENDIX: WHY THE ATTACK ON FOSSIL FUELS IS UNSCIENTIFIC

THE "SKEPTIC" SMEAR BY ERIC DENNIS

The most frustrating thing about being a scientist skeptical of catastrophic global warming is that the other side is continually distorting what I am skeptical of.

In his celebrated 2012 *New York Review of Books* article "Why the Global Warming Skeptics Are Wrong," economist William Nordhaus presents six questions that the legitimacy of global warming skepticism allegedly rests on: [89]

1. Is the planet in fact warming?

2. Are human influences an important contributor to warming?

3. Is carbon dioxide a pollutant?

4. Are we seeing a regime of fear for skeptical climate scientists?

5. Are the views of mainstream climate scientists driven primarily by the desire for financial gain?

6. Is it true that more carbon dioxide and additional warming will be beneficial?

Since the answers to these questions are allegedly yes, yes, yes and no, no, no, it's case closed, says Nordhaus.

Except that he is attacking a straw man. Scientists (or non-scientists) who are "skeptics" are skeptical of catastrophic global warming — not warming or human-caused warming as such. So much for 1 and 2. We refuse to label CO_2 a "pollutant" because it is essential to life and because we do not believe it has the claimed catastrophic impact. So much for 3. And since 4-6 don't pertain to the scientific issue of catastrophic warming, so much for them, as well.

The object of our skepticism, *catastrophic* global warming, means warming caused by greenhouse gases that would so dramatically heat up the earth that despite the proven climate adaptability of hydrocarbon-powered civilization, populations the world over would experience impoverishment, mass suffering, and death.[90]

Why are we skeptical of this claim? Because there is radically insufficient evidence for it.

This may seem implausible, because the news media bombard us with stories of new studies, new findings, new models, new international summits allegedly confirming catastrophic global warming. But what these stories leave out is the evidential status of these developments—what any given study or model actually proves. And the answer is, little to nothing, because the present ability of scientists to understand, model, and predict the climate is far, far lower than we are led to believe.

To say that modeling the climate for long-term predictions is difficult given the current state of climate science is like saying that it would be difficult for your five-year-old son to build a 400 horsepower car from re-purposed Toys 'R' Us purchases. Imagine that he comes to you with pages and pages of plans he's sketched out in crayon. The "car" will cost $22,827.35 worth of toys.

Why wouldn't you reach for your credit card? Is that because you're against teaching kids engineering? Is it because his sworn enemy, your daughter, is paying you off? Or perhaps it's because this project is obviously beyond the capability of a five-year-old, and that his crayon schematics don't offer convincing evidence that he is in fact the kind of once-in-a-generation prodigy who could somehow pull it off.

If one understands how monumental an undertaking it would be to produce a sound climate model, one can see that today's climate modelers are making assertions no less implausible than our five-year old's fantasy.

In physics it is generally possible to exactly predict the behavior of systems involving two independent bodies, whether planets interacting through gravity or elementary particles through the electromagnetic field. More bodies means no exact solution to the dynamical equations and a zoo of different approximations, usually requiring computational simulation, which takes more and more time as the number of bodies being simulated increases. Indeed the computation time generally grows exponentially with the number of bodies.

The global climate system comprises an astronomical number (at least billions) of effectively independent "bodies," which is to say of isolatable, relatively uniform chunks of air, ocean, and earth. Their interactions span the complexity spectrum, from the mechanical push-and-pull of an ocean current to the lesser-known dynamics of cloud formation to intricate, biological mechanisms like plant growth and respiration that have evolved over billions of years.

Solving this kind of complex system is outside the realm of controlled approximations and reasonable estimates. It's in the realm of random stabs, on any objective assessment of our current scientific powers. Since attempts to model this system are the basis of claims for catastrophic global warming, the evidence we need to consider pertains to whether or not such models are capturing enough of the detailed mess of forces that actually drives the climate.

Many different climate processes affect the energy balance between the earth and outer-space, and thus affect temperatures on the Earth. One such process is the greenhouse effect, by which CO_2 and other gases trap some extra solar energy in the atmosphere and convert it into heat. It is widely acknowledged that the CO_2-linked greenhouse effect itself can produce only a modest warming going forward because the incremental warming produced by each extra liter of CO_2 gets smaller and smaller as more CO_2 is added.

The catastrophist projections are based on the idea that this modest warming will trigger an entirely separate set of feedback mechanisms that will multiply the warming many times.[91] For instance: warming is projected to increase ambient levels of water vapor, itself a greenhouse gas; melting ice will expose more earth or open water, which tend to absorb more solar energy as heat; temperature-linked changes in cloud patterns affect how much solar energy gets reflected back to space or back to the Earth.

There are also negative feedbacks, meaning processes that come into play due to warming, or to CO_2 increases, that wind up counteracting that warming. Examples include enhanced re-radiation of energy back into space at higher temperatures, increased absorption of CO_2 into the oceans, and increased quantities of organic matter capturing CO_2. Indeed some supposedly positive feedbacks, like certain cloud effects, may turn out actually to be negative ones.[92]

Moreover, nature does not simply provide us with a list of all the relevant feedbacks, or climate processes in general. There is no systematic procedure by which the set of processes included in current climate models are picked out from the catalogue of all possible such processes. The procedure is simply for modelers to engage their own imaginations, given our current knowledge, to conceive possible effects and gather evidence to confirm or falsify them.

How many known ones have been intentionally discarded due to a lack of knowledge and evidence about how to incorporate them? How many have just not been thought of to date?

In a certain sense, this is the nature of any scientific theory. But this is why such theories have to produce specific, detailed predictions, confirmed by observation, to show that they have captured the relevant causal factors. Apart from this, there is a lot of room here for the ultimate outcome of the models to be controlled by ideological predispositions—like that, of all the underlying drivers, the decisive one just happens to be CO_2, the one with a clear link to the functioning of modern, industrial capitalism.

What would be a rational response when your five-year-old car enthusiast presents you with his crayon plans, protesting that he's also proven his case by putting together a scale model in Legos? First you might point out that while his plans are impressive for a boy his age, it's rarely the case that reality works out just like *a priori* plans and models suggest.

Rather than setting him loose at toysrus.com with your credit card, you might suggest he start off with a scaled-down project, like an RC kit. Then, if that's a success, maybe an introduction to simple wood and then metal work. As he gets older and proves himself at each stage, he could move on to machine shop projects, welding, and an apprenticeship with a real car mechanic.

This kind of demonstrated, step-by-step progress is how legitimate inventions, and inventors, are made. At the end of the process, they no longer agitate for sizable investments on the basis of their original crayon plans.

And such demonstrated, step-by-step progress is exactly what a reasonable person ought to demand from the global warming catastrophists. Not mere simulations, generated by model code that they control and have played with for years. Since the odds are so small, *a priori*, that they have actually cracked the excruciatingly complicated problem of global climate prediction, we need dramatic positive evidence. Lesser evidence is powerless to overcome the overwhelming odds against being able to delicately sort out the mess of climate drivers and feedbacks.

The catastrophists need to demonstrate their methodology by applying it to smaller problems whose outcomes we don't have to wait a century for. They need to derive unambiguous, detailed predictions for these outcomes and see them borne out. By "detailed" I mean predictions of not just a single number, like a cumulative warming trend, that could just be accidentally correct— and they're not even getting predictions on these simpler metrics right.[93] I mean predictions of a more intricate, unaccidental nature.

For instance, climate models predict a detailed pattern of warming that occurs at different rates in different parts of the globe and, importantly, at different altitudes in the atmosphere. But when we look in actual climate data for the specific, altitude-dependent warming signature produced by these models, we find something entirely different.[94]

And that's only half the problem. Before we can test models, we need this historical climate data to be accurate in order for the comparison to mean anything. Even for the one central climate variable, global average temperature, the reconstructed data is fraught with uncertainties and scientific misconduct.[95]

What always has to be kept in mind on these issues is:

- The massive complexity of the problem the catastrophist modelers are claiming to have solved relative to the current state of climate science.

- What this implies about the onus of proof.

Their claim is to have accomplished a scientific miracle with tools that by any reasonable analysis are far from capable of the task.

Absent shocking evidence of success on their part, the conclusion to draw is not: catastrophic global warming has just moderate odds of occurring. The conclusion is that these models bear as much relationship to reality as your son's crayon plans bear to a real car. And suggestions about how to transform the entire world economy based on these models should be treated accordingly.

HOW BAD SCIENCE BECOMES COMMON KNOWLEDGE
BY ERIC DENNIS

To read the popular media's account of climate science, it is a certainty that burning fossil fuels is causing an unprecedented and catastrophic warming of the planet. The volume of such claims is so vast that those skeptical of catastrophic warming are often viewed as conspiracy theorists, believing that scientists and the media have formed a secret cabal to foist falsehoods on the public.

But the case for being skeptical of catastrophic warming—and, more broadly, many popular scientific assertions—has nothing to do with conspiracy theories. It is based on knowledge of the mechanism by which new scientific ideas are evaluated and spread by non-experts, who are prone to choose winners and losers on the basis of congenial political ideology rather than scientific merit.

Case 1: Aidan Dwyer's "Breakthrough"

A 2012 episode in the science and tech media illustrates this mechanism. Gizmodo,[96] The Atlantic Wire[97], and Popular Science[98] all lauded a new "breakthrough" at the hands of a 13-year-old "genius," Aidan Dwyer, first recognized by the American Museum of Natural History with its Young Naturalist Award.[99]

His insight? A "super-efficient solar array" differing from standard arrays in one respect: the arrangement of individual solar cells at various random-looking angles according to a specific mathematical pattern (the Fibonacci sequence) that characterizes the leaves and branches of certain trees.

By all accounts, Aidan Dwyer is a bright, well-meaning boy. But this proposal makes no sense, and he has ultimately been ill-served by the adults lauding it. The normal configuration of solar panels has each cell oriented at the angle yielding optimal total exposure to the sun's day-long path in the sky. Each cell is either oriented at that one optimal angle or at a sub-optimal angle producing less output power —and mimicking a tree is far from optimal.

But notice that the *narrative* is optimal to two generations of media members steeped in "green" ideology: an innocent prodigy, influenced by the beauty and wisdom of nature, imposes natural order on brute technology to prove the viability of green energy. And so those media members, lacking any particular expertise on solar panels, ran with it.

Aidan Dwyer would never have received the same acclaim had he, say, conducted an experiment in his family's garage leading him to claim the discovery of a new chemical agent for fracing. Can anyone imagine that the most prominent natural history museum in the country would then give him an award and the media would trumpet the arrival of a budding genius in the field of energy research? Of course not.

This episode is important because it shows, in microcosm, how much of what passes for common knowledge comes to be. From the vast well of concrete events and ideas in science and technology, certain ones are picked up and amplified while others are discarded by the network of influencers and disseminators— from government bureaucrats awarding the grants that academic science lives on, to the mainstream media publishing what it regards as the most important findings.

The vast, vast majority of the network is by necessity non-expert on any given topic. In an advanced, division-of-labor society, there is a division of scientific expertise. That is a good thing, as it enables a staggering total of knowledge to be discovered and applied throughout society. But there is an ever-present hazard of loud or numerous non-experts promoting views as certainties because those views fit their political ideologies.

Case 2: Michael Mann's "Hockey Stick"

And that is exactly what has happened with global warming. For example, when we hear of vast numbers of scientists endorsing Michael Mann's famous "hockey stick" graph—the rhetorical star of Al Gore's "An Inconvenient Truth"—what we don't hear is that the vast, vast majority of them never sought access to the specific data and algorithms claimed to support it (much of these have been actively withheld from the scientific community at large).[100]

They did not independently evaluate either Mann's claims or the specific, technical objections raised against them by a few critics who were able to wrest those data and algorithms from Mann's clenched fist over a period of years. Neither had the scientific media performed any independent, critical review when reporting on such issues for over a decade, most of them simply not being equipped to do so.

From the perspective of those among the green-leaning media who actually are equipped by this point to verify reports of serious flaws in Mann's approach, why exert all that effort with the hope of merely confirming what is already an ideological pillar, when a positive result would be superfluous and a negative one would be, at best, ominously confusing? This attitude is in fact embraced by climatologists at the highest levels.

After a critic asked renowned climatologist Phil Jones to release the raw data from which he has generated one of the primary historical records of global temperature, Jones's famous response was "Why should I make the data available to you, when your aim is to try and find something wrong with it?"[101]

It is now generally acknowledged that Michael Mann's original claims about a precipitous acceleration in global warming around the advent of industrialization were founded on a broken methodology. As shown originally by two Canadian researchers, [102] and verified by a U.S. Senate-appointed expert panel of independent statisticians, the technique indicates precipitous warming, whether fed with actual climate data or with simulated data designed to lack any underlying trend at all. [103]

Yet it was not until five years after Mann's original publication—and after the hockey stick graph was immortalized by the ostensible cream of international climate expertise at the Intergovernmental Panel on Climate Change (IPCC)—that the broken parts under its hood were first identified in a scientific journal. And this was accomplished not by any of Mann's colleagues at Penn State, nor any of his many co-authors, peer-reviewers, or IPCC editors. It was accomplished by a mathematically savvy mining consultant, Steve McIntyre, and an economist, Ross McKitrick, who both took it up essentially as a hobby, receiving not one of the billions of dollars in government climatology funding funneled to academic researchers. [104]

The same basic mechanism that made Aidan Dwyer a star has, on a different level, made Michael Mann a star. The primary difference is the level of technical sophistication—a level in the latter case just high enough to be dangerous in a realm where even expert statisticians (which climatologists are not) have to be on guard against inconspicuous but critical errors.

Enthralling your average climatologist requires something subtler than the mathematics of branch growth patterns, something more like Michael Mann's novel statistical technique to extract imperceptible trends from a hodgepodge of tree ring and ice core measurements that seem to imply a dangerous acceleration in warming circa 1900 (the "hockey stick" graph), hence an ideologically convenient fatal flaw in industrial capitalism.

Note that this is especially dangerous in a field such as climatology, where there are zero experts who can accurately predict how various important but poorly understood factors will come together to drive the climate. This is a field ripe for ideological grant-givers to make superstars out of intellectually immodest mediocrities.

And just as Aidan Dwyer's celebrity carries on despite clear technical refutation, so the global warming movement carries on despite the hockey stick having been split asunder by clear proof of the inherent hockey-stick bias in Mann's statistical technique.

Disseminating Good Science

None of this implies any cognitive determinism for climatologists or pop-science consumers sharing a common world-view. Each one is free to think for himself, to gather new data perhaps through alternative networks, and to assess the totality of evidence available to him. But such tasks require an effort whose mark many want to display without going to the trouble of exerting it, as is demonstrably the case with the denizens of the global warming movement. So arises the widespread belief that we're facing a climate crisis, that the "green" technology is out there to replace fossil fuels, and that it's just a matter of getting the right set of bright young kids working in the right direction.

To some extent the intellectual division of labor will always mean that there is no guarantee against large-scale, ideologically driven mistakes gaining wide currency. However this is especially probable in the present, monolithic system of government-funded basic research, where bureaucrats carelessly appropriate money they didn't earn towards projects whose benefits they won't receive, inspired by ideology-laden fads whose underlying accuracy they are not particularly concerned with.

The elimination of the profit-motive does not banish individuals' pursuit of their own interests; it redirects that pursuit away from honest value creation and into a distorted, unspoken realm of indirect benefits and cynical power bartering among appropriators whose one common goal is the expansion of their appropriation stream.

What we need is to restore the profit-motive in a system of free individuals, pursuing their own goals openly with their own wealth. It is said that such a system will stifle visionary thinkers whose ideas are too long-range to make a quick buck. But this is just a smokescreen obscuring what profit-and-loss in a system of well-defined property rights—profits whose range is much longer than the next election—are uniquely capable of factoring into such investment decisions: the inescapable trade-off between the revolutionary power of basic research and the probability of concrete benefits flowing from it.

Large Stakes

What's at stake is the lives of billions of people in the present and future. Their lives depend on access to industrial technology that scientifically illiterate politicians around the world are subjecting to the ransom of their regulations and controls. Ransom letters are delivered to us daily from the op-eds, the articles, and the talking heads educating us about thousands of experts that have all verified the coming of an apocalypse against which our only savior, conveniently, is more climatology research funding and more concentrated political power.

ABOUT THE AUTHOR

Alex Epstein, President of the Center for Industrial Progress, is a leading philosopher of energy, specializing in the moral case for fossil fuels and nuclear power. His writings on energy have been published in The Wall Street Journal, Forbes, and Investor's Business Daily, and he has debated Greenpeace, Occupy Wall Street, and 350.org.

To inquire about bringing Mr. Epstein to speak at your company or group, email support@industrialprogress.net.

ENDNOTES

[1] Indur M. Goklany, "Humanity Unbound: How Fossil Fuels Saved Humanity From Nature and Nature From Humanity," *CATO Policy Analysis No. 715* (Washington D.C.: CATO Institute, December 20, 2012*)*, 1; 7, accessed February 15, 2013, http://www.cato.org/publications/policy-analysis/humanity-unbound-how-fossil-fuels-saved-humanity-nature-nature-humanity.

[2] Goklany, "Humanity Unbound: How Fossil Fuels Saved Humanity From Nature and Nature From Humanity," 6, 8, 13, 14, 25.

[3] BP Statistical Review of World Energy, June 2012.

[4] World Energy Outlook 2012, Chapter 18, accessed February 16, 2013, http://www.worldenergyoutlook.org/media/weowebsite/energydevelopment/2012updates/Measuringprogresstowardsenergyforall_WEO2012.pdf.

[5] Alex Epstein, "Four Dirty Secrets About Clean Energy," *Fox News,* June 3, 2011, accessed February 16, 2013, http://www.foxnews.com/opinion/2011/06/03/four-dirty-secrets-about-clean-energy/.

[6] Anthony Watts, "Saturday Silliness—Josh's Wind Energy Fact Sheet—Global Wind Power 'to the Nearest Whole Number,'" *Watts Up With That?,* 2007, accessed February 16, 2013, http://wattsupwiththat.com/2012/03/10/saturday-silliness-joshs-wind-energy-fact-sheet-global-wind-power-to-the-nearest-whole-number/.

[7] Stefan Nicola and Tino Andresen, "Merkel's Green Shift Forces Germany to Burn More Coal," *Bloomberg, a*ccessed February 17, 2013, http://www.bloomberg.com/news/2012-08-19/merkel-s-green-shift-forces-germany-to-burn-more-coal-energy.html.

[8] Federal Statistical Office of Germany, "Gross Electricity Production in Germany 2010 - 2012," accessed February 15, 2013, https://www.destatis.de/EN/FactsFigures/EconomicSectors/Energy/Production/Tables/GrossElectricityProduction.html.

[9] Ami Sedghi and John Burn-Murdoch, "Unemployment in Europe: get the figures for every country," *The Guardian,* accessed February 16, 2013, http://www.guardian.co.uk/news/datablog/2012/oct/31/europe-unemployment-rate-by-country-eurozone#data.

[10] Goklany, "Humanity Unbound: How Fossil Fuels Saved Humanity From Nature and Nature From Humanity," 7.

[11] Ibid., 20.

[12] Caroline, "Panel Plant Pollution," *Things Worse Than Nuclear Power Blog,* accessed February 16, 2013, http://www.thingsworsethannuclearpower.com/2012/03/panel-plant-pollution.html.

[13] Caroline, "The Real Waste Problem, Solar Edition," *Things Worse Than Nuclear Power Blog,* accessed February 16, 2013, http://www.thingsworsethannuclearpower.com/2012/09/the-real-waste-problem-solar-edition.html.

[14] "Polycrystalline Thin-Film Materials and Devices R&D," *National Renewable Energy Laboratory*, 2011, accessed February 16, 2013, http://www.nrel.gov/pv/thinfilm.html.

[15] Simon Parry and Ed Douglas, "In China, the true cost of Britain's clean, green wind power experiment: Pollution on a disastrous scale," *Daily Mail*, January 26, 2011, accessed February 16, 2013, http://www.dailymail.co.uk/home/moslive/article-1350811/In-China-true-cost-Britains-clean-green-wind-power-experiment-Pollution-disastrous-scale.html.

[16] Pierre Desrochers and Hiroko Shimizu, *The Locavore's Dilemma: In Praise of the 10,000-mile Diet,* (New York: PublicAffairs, 2012) 5, 14.

[17] Keith H. Lockitch, "Climate Vulnerability and the Indispensable Value of Industrial Capitalism", *Energy & Environment* 20 (2009), accessed February 16, 2013, http://heartland.org/sites/all/modules/custom/heartland_migration/files/pdfs/25905.pdf.

[18] Indur M. Goklany, "Wealth and Safety: The Amazing Decline in Deaths from Extreme Weather in an Era of Global Warming, 1900—2010", *Reason Foundation Policy Study No. 393*, September 2011, accessed February 17, 2013, http://reason.org/files/deaths_from_extreme_weather_1900_2010.pdf.

[19] Christian Parenti, "Climate Action Opponents Are Ensuring the Outcome They Claim to Oppose: Big Government" *ThinkProgress Blog,* January 31, 2011, accessed February 24, 2013, http://thinkprogress.org/climate/2012/01/31/414155/climate-action-big-government/?mobile=nc.

[20] U.S. Department of Health and Human Services, Centers for Disease Control and Prevention, National Center for Health Statistics, "Health, United States, 2011," 108, accessed February 24, 2013, http://www.cdc.gov/nchs/data/hus/hus11.pdf#022.

[21] Parenti, "Climate Action Opponents Are Ensuring the Outcome They Claim to Oppose: Big Government."

[22] George Reisman, "Global Warming Is Not a Threat but the Environmentalist Response to It Is," *George Reisman's Blog on Economics, Politics, Society and Culture,* May 30, 2007, accessed February 15, 2013, http://georgereismansblog.blogspot.com/2007/05/global-warming-is-not-threat-but.html.

[23] L., Emily, "Robert F. Kennedy, Jr: 'Coal is Crime,'" *Care2,* May 8, 2012, accessed February 17, 2013, http://www.care2.com/causes/robert-f-kennedy-jr-coal-is-crime.html.

[24] UNICEF, *Progress on Drinking Water and Sanitation 2012 Update*, accessed Feb. 15, 2013, http://www.unicef.org/media/files/JMPreport2012.pdf.

[25] Ibid.

[26] Kathleen Hartnett White, "EPA's Pretence of Science: Regulating Phantom Risks," *Texas Public Policy Foundation*, May 2012, accessed Feb 16, 2013, http://www.texaspolicy.com/sites/default/files/documents/epa-pretense-of-science-acee-kathleen-hartnett-white.pdf.

[27] Bernard L. Cohen, *The Nuclear Energy Option* (New York: Plenum Press, 1990), Ch. 5, http://www.phyast.pitt.edu/~blc/book/chapter5.html.

[28] Parineeta Dandekar, "Where Rivers Run Free—Policy Tools to Protect Free-Flowing Rivers," *International Rivers,* accessed February 17, 2013, http://www.internationalrivers.org/resources/where-rivers-run-free-1670.

[29] Al Gore, "A Generational Challenge to Repower America," July, 17, 2008, accessed February 15, 2013, http://blog.algore.com/2008/07/a_generational_challenge_to_re.html.

[30] BP Statistical Review of World Energy, June 2012, 41, http://www.bp.com/assets/bp_internet/globalbp/globalbp_uk_english/reports_and_publications/statistical_energy_review_2011/STAGING/local_assets/pdf/statistical_review_of_world_energy_full_report_2012.pdf.

[31] John M. Broder, "Obama Affirms Climate Change Goals," *The New York Times*, November 18, 2008, accessed February 17, 2013, http://www.nytimes.com/2008/11/19/us/politics/19climate.html.

[32] U.S. Department of Health and Human Services, Centers for Disease Control and Prevention, National Center for Health Statistics, "Health, United States, 2011," 108.

[33] Goklany, "Humanity Unbound: How Fossil Fuels Saved Humanity From Nature and Nature From Humanity," 3, 7, 24, 27.

[34] Federal Ministry of Economy and Technology, Zahlen und Fakten - Energiedaten, accessed February 16, 2013. http://www.bmwi.de/DE/Themen/Energie/energiedaten.html.

[35] Robert Bryce, "Coal Hard Facts," in *Power Hungry: The Myths of "Green" Energy and the Real Fuels of the Future* (New York: Perseus Books Group, 2010).

[36] UNICEF, *Progress on Drinking Water and Sanitation 2012 Update.*

[37] Goklany, "Wealth and Safety."

[38] "Fukishima Accident 2011," World Nuclear Association, last modified January 10, 2013, accessed February 24, 2013, http://www.world-nuclear.org/info/fukushima_accident_inf129.html.

[39] U.S. Nuclear Regulatory Commission, "Backgrounder on the Three Mile Island Accident," last modified February 11, 2013, accessed February 24, 2012, http://www.nrc.gov/reading-rm/doc-collections/fact-sheets/3mile-isle.html.

[40] "Chernobyl Accident 1986," World Nuclear Association, last modified December 2012, accessed February 24, 2013, http://www.world-nuclear.org/info/chernobyl/inf07.html.

[41] Jason Morgan, "Energy Density and Waste Comparison of Energy Production," *Nuclear Fissionary*, June 9, 2010, accessed February 24, 2013, http://nuclearfissionary.com/2010/06/09/energy-density-and-waste-comparison-of-energy-production/.

[42] Bernard L. Cohen, "Nuclear Power in France," World Nuclear Association, April 2012, http://www.world-nuclear.org/info/default.aspx?id=406&terms=world%20energy%20production.

[43] BP Statistical Review of World Energy, June 2012, 41.

[44] Sue Halpern, "Who Was Steve Jobs?" *The New York Review of Books*, January 12, 2012, http://www.nybooks.com/articles/archives/2012/jan/12/who-was-steve-jobs/.

[45] Dustin Mulvaney, Vicki Bolam, Monica Cendejas, Sheila Davis, Lauren Ornelas, Simon Kim, Serena Mau, William Rowan, Esperanza Sanz, Peter Satre, Ananth Sridhar, Dean Young, "Toward a Just and Sustainable Solar Energy Industry," A Silicon Valley Toxics Coalition White Paper, January 14, 2009, accessed February 16, 2013, http://svtc.org/wp-content/uploads/Silicon_Valley_Toxics_Coalition_-_Toward_a_Just_and_Sust.pdf.

[46] Goklany, "Humanity Unbound: How Fossil Fuels Saved Humanity From Nature and Nature From Humanity," 3, 4, 6, 7, 24.

[47] U.S. Forest Service, "1979: Snail Darter Exemption Case," last modified November 24, 2008, accessed February 16, 2013, http://www.foresthistory.org/ASPNET/Policy/northern_spotted_owl/1979owl.snaildarter.aspx

[48] Caroline, "The Real Waste Problem, Solar Edition."

[49] Steve Pociask and Joseph P. Fuhr Jr., "Progress Denied: A Study on the Potential Economic Impact of Permitting Challenges Facing Proposed Energy Project," U.S. Chamber of Commerce—Project No Project, 32, 33, 37, 45, 46, 53, 75, accessed Feb 15, 2013, http://www.projectnoproject.com.

[50] Paul Ehrlich and Anne Ehrlich, "Too Many People, Too Much Consumption," *Yale Environment 360 Blog*, August 4, 2008, accessed February 15, 2012, http://e360.yale.edu/feature/too_many_people_too_much_consumption/2041/.

[51] Nicholar Ballasy, "White House Science Czar Says He Would Use 'Free Market' to De-Develop the United States," *CNS News*, September 16, 2010, accessed February 15, 2013, http://cnsnews.com/node/75388.

[52] Ted Turner, "What Liberals Say," *Accuracy in Media*, accessed February 15, 2013, http://www.aim.org/wls/five-percent-of-the-present-population-would-be-ideal/.

[53] Paul Ciotti, "Fear of Fusion: What If It Works?" *Los Angeles Times*, April 19, 1989, accessed February 15, 2013, http://articles.latimes.com/1989-04-19/news/vw-2042_1_fusion-uc-berkeley-inexhaustible.

[54] "'Climate-gate': Beyond the embarrassment," *Science News*, December 12, 2009, accessed February 15, 2013, http://www.sciencenews.org/view/generic/id/50711/description/.

[55] "Why destroy America's foremost wildlife refuge for less oil than we consume in a single year?" *Natural Resource Defense Council*, The Arctic National Wildlife Refuge, last modified March 10, 2005, accessed February 24, 2013, http://www.nrdc.org/land/wilderness/arcticrefuge/facts1.asp.

[56] James Delaney, "Rousseau, Jean-Jacques," *Internet Encyclopedia of Philosophy*, 2005, accessed February 15, 2013, http://www.iep.utm.edu/rousseau/.

[57] Eric Morris, "From Horse Power to Horsepower," *Access Magazine* (30), University of Califormia Transportation Center, acessed February 16, 2013, http://www.uctc.net/access/30/Access%2030%20-%2002%20-%20Horse%20Power.pdf.

[58] "McKibben vs. Epstein," accessed February 17, 2013, http://fossilfueldebate.com/.

[59] David Rose, "Global Warming Stopped 16 Years Ago, Reveals Met Office Report Quietly Released... and Here Is the Chart to Prove It," *Mail Online*, October 13, 2012, accessed February 24, 2012, http://www.dailymail.co.uk/sciencetech/article-2217286/Global-warming-stopped-16-years-ago-reveals-Met-Office-report-quietly-released--chart-prove-it.html.

[60] Goklany, "Humanity Unbound: How Fossil Fuels Saved Humanity From Nature and Nature From Humanity," 20.

[61] BP Statistical Review of World Energy, June 2012.

[62] "McKibben vs. Epstein."

[63] "McKibben vs. Epstein."

[64] Nicola and Andresen, "Merkel's Green Shift Forces Germany to Burn More Coal."

[65] "McKibben vs. Epstein."

[66]Jan Fleischhauer and Alexander Neubacher, "German Energy Agency Chief: 'We'll Need Conventional Power Plants until 2050," *Spiegel Online International*, November 15, 2012, accessed February 16, 2013, http://www.spiegel.de/international/germany/german-energy-expert-argues-against-subsidies-for-solar-power-a-866996.html.

[67] Devon Swezey, "Doing the Math: Comparing Germany's Solar Industry to Japan's Fukushima Reactor," *The Breakthrough*, March 23, 2011, accessed February 17, 2013, http://thebreakthrough.org/archive/doing_the_math_comparing_germa.

[68] "McKibben vs. Epstein."

[69] Dr.-Ing. Günter Keil, "Exaggerating, euphemizing, ignoring, concealing," and "Wind power—supply according to the weather," in *Germany's Energy Supply Transformation Has Already Failed*, December 2011, accesssed February 16, 2013. http://www.eike-klima-energie.eu/uploads/media/2012_01_09_EIKE_Germa_energy_turnaround_english.pdf.

[70] National Research Council, *Hidden Costs of Energy: Unpriced Consequences of Energy Production and Use* (Washington, DC: The National Academies Press, 2010), http://www.nap.edu/catalog.php?record_id=12794.

[71] Paul Krugman, "Here Comes Solar Energy." *The New York Times*, November 6, 2011, http://www.nytimes.com/2011/11/07/opinion/krugman-here-comes-solar-energy.html.

[72] *Would You Give Up The Internet For 1 Million Dollars?* YouTube Video, 5:10, posted by "TFASvideo" 2011, http://youtu.be/0FB0EhPM_M4..

[73] Ibid.

[74] Chris Landsea, "Why don't we try to destroy tropical cyclones by nuking them?" U.S. Deparment of Commerce, National Oceanic and Atmospheric Administration, Hurricane Research Division, Tropical Cyclones Myth Page, last modified 2006, accessed February 15, 2013, http://www.aoml.noaa.gov/hrd/tcfaq/C5c.html.

[75] U.S. Department of Justice, Office of Justice Programs, Bureau of Justice Statistics, "Local Police," accessed February 15, 2013, http://bjs.ojp.usdoj.gov/index.cfm?ty=tp&tid=71.

[76] U.S. Fire Administration, "National Fire Protection Association Estimates," accessed February 15, 2013, http://www.usfa.fema.gov/statistics/estimates/nfpa/index.shtm.

[77] Roger Mitchell, "Tons of Oil Seeps into Gulf of Mexico Each Year," NASA Earth Observatory, accessed February 15, 2013, http://earthobservatory.nasa.gov/Newsroom/view.php?id=20863.

[78] Simon Parry, "In China, the true cost of Britain's clean, green wind power experiment: Pollution on a disastrous scale," Mail Online, January 26, 2011, accessed May 10, 2013, http://www.dailymail.co.uk/home/moslive/article-1350811/In-China-true-cost-Britains-clean-green-wind-power-experiment-Pollution-disastrous-scale.html.

[79] ExxonMobil, accessed February 24, 2013, http://www.exxonmobil.com/Corporate/.

[80] Shell, accessed February 24, 2013, http://www.shell.com/.

[81] Chevron, accessed February 24, 2013, http://www.chevron.com/.

[82] "I Love Fossil Fuels," Facebook Page, accessed February 17, 2013, http://www.facebook.com/ILoveFossilFuels/photos_stream.

[83] Chevron, "We Agree, Do You?" accessed February 16, 2013, www.chevron.com/weagree/.

[84] Food and Agriculture Organization of the United Nations, *The State of Food Insecurity in the World 2012*, Annex 1, accessed Feb 16, 2013, http://www.fao.org/docrep/016/i3027e/i3027e06.pdf.

[85] Daniel Yergin, "From Shortage to Surplus: The Age of Gasoline," in *The Prize: The Epic Quest for Oil, Money and Power*. (New York: Free Press, 1991), 194.

[86] UNICEF, *Progress on Drinking Water and Sanitation 2012 Update*.

[87] U.S. Energy Information, International Energy Statistics, Coal, Consumption, accessed February 17, 2013, http://www.eia.gov/cfapps/ipdbproject/iedindex3.cfm?tid=1&pid=1&aid=2&cid=regions&syid=1990&eyid=2011&unit=TST

[88] The World Bank, "Life expectancy at birth, total (years)," accessed February 17, 2013, http://data.worldbank.org/indicator/SP.DYN.LE00.IN

[89] William Nordhaus, "Why the Global Warming Skeptics Are Wrong," *The New York Review of Books*, February 22, 2012, accessed February 16, 2013, http://www.nybooks.com/articles/archives/2012/mar/22/why-global-warming-skeptics-are-wrong/?pagination=false.

[90] Alex Epstein, "How Capitalism Makes Catastrophes Non-Catastrophic," *MasterResource Blog*, February 12, 2012, accessed February 16, 2013. http://www.masterresource.org/2012/02/how-capitalism-makes-catastrophes-non-catastrophic/.

[91] National Research Council, *Understanding Climate Change Feedbacks* (Washington, DC: The National Academies Press, 2003), http://www.nap.edu/catalog.php?record_id=10850.

[92] Spencer, R. W., W. D. Braswell, J. R. Christy, and J. Hnilo. "Cloud and radiation budget changes associated with tropical intraseasonal oscillations," *Geophysical Research Letters* 34 (2007), L15707, accessed February 16, 2013, doi: 10.1029/2007GL029698.

[93] Nir Shaviv, "On IPCCs exaggerated climate sensitivity and the emperor's new clothes," *ScienceBits Blog*, January 9, 2012, accessed February 16, 2013, http://www.sciencebits.com/IPCC_nowarming.

[94] Douglass, D. H., Christy, J. R., Pearson, B. D. and Singer, S. F. "A comparison of tropical temperature trends with model predictions," International Journal of Climatology 28 (2007): 1693−1701, accessed February 16, 2013. doi: 10.1002/joc.1651.

[95] Dr. Eric Dennis, "How Bad Science Becomes Common Knowledge: Two Case Studies (solar and climate change)," *MasterResource Blog*, January 17, 2012, accessed February 16, 2013, http://www.masterresource.org/2012/01/bad-climate-science-common-knowledge/.

[96] Brent Rose, "Genius 13-Year-Old Has a Solar Power Breakthrough." *Gizmodo*, April 19, 2011, accessed February 24, 2013, http://gizmodo.com/5832557/genius-13+year+old-has-a-solar-power-breakthrough.

[97] Adam Martin, "13-Year-Old Looks at Trees, Makes Solar Power Breakthrough," *The Atlantic Wire*, August 19, 2011, accessed February 24, 2013, http://www.theatlanticwire.com/technology/2011/08/13-year-old-looks-trees-makes-solar-power-breakthrough/41486/.

[98] Rebecca Boyle,"13-Year-Old Designs Super-Efficient Solar Array Based on the Fibonacci Sequence," *Popular Science*, August 19, 2011, accessed February 24, 2013, http://www.popsci.com/technology/article/2011-08/13-year-old-designs-breakthrough-solar-array-based-fibonacci-sequence.

[99] American Museum of Natural History, "American Museum of Natural History Young Naturalist Awards for Student Scientists," accessed February 16, 2013, http://www.amnh.org/about-us/press-center/2011-young-naturalist-award-winners.

[100] Michael Le Page, "Climate myths: The 'hockey stick' graph has been proven wrong," NewScientist, last modified September 2009, accessed February 24, 2013, http://www.newscientist.com/article/dn11646-climate-myths-the-hockey-stick-graph-has-been-proven-wrong.html.

[101] Patrick J. Michaels, "The Dog Ate Global Warming," National Review Online, September 23, 2009, accessed February 24, 2013, http://www.nationalreview.com/articles/228291/dog-ate-global-warming/patrick-j-michaels?pg=1.

[102] Ross McKitrick, "What is the 'Hockey Stick' Debate About?" *University of Guelph*, Department of Economics, April 4, 2005, accessed February 15, 2013, http://www.uoguelph.ca/%7Ermckitri/research/McKitrick-hockeystick.pdf.

[103] E. Wegman, D. Scott, and Y. Said, "Ad Hoc Committee Report on the 'Hockey Stick' Global Climate Reconstruction," *Science and Public Policy Institute*, April 26, 2010, accessed February 15, 2015, http://scienceandpublicpolicy.org/reprint/ad_hoc_report.html.

[104] Ibid.

CPSIA information can be obtained
at www.ICGtesting.com
Printed in the USA
BVHW072306080622
639212BV00001B/61

9 780989 344807